ELECTRONIC ASSEMBLY

**THE FUNDAMENTALS SERIES
FOR ELECTRONIC TECHNOLOGY**

Electronic Assembly: Concepts and Experimentation

Fundamental DC/AC Circuits: Concepts and Experimentation

Fundamental Electronic Devices: Concepts and Experimentation,
   Second Edition

Digital Electronics: Theory and Experimentation

Microprocessor Technology: Theory, Experimentation and
   Troubleshooting

ELECTRONIC ASSEMBLY
Concepts and Experimentation

Fredrick W. Hughes

Electronics Training Consultant

THE FUNDAMENTALS SERIES FOR ELECTRONIC TECHNOLOGY

PRENTICE HALL CAREER & TECHNOLOGY
Upper Saddle River, NJ 07458

Library of Congress Cataloging-in-Publication Data

Hughes, Fredrick W.
   Electronic assembly : concepts and experimentation / Fredrick W.
Hughes.
      p. cm.
   Includes index.
   ISBN 0-13-249731-X
   1. Electronic apparatus and appliances—Design and construction.
TK7870.H85 1992
621.381—dc20   91-33796
                                           CIP

Dedication

This book is dedicated to Mr. Robert Roberts whose persistence in the field of education and the training of Electronics Assembly workers has earned him a well deserved retirement. His invaluable assistance and cooperation was most appreciated. Good luck BOB!

Also special recognition is given to Mr. Gil Patterson for the cover photo and his assistance in photography and the moral support he provided. Okay Gil, it's time for you to publish!

Special thanks to my wife Roberta and son Jeffery for their assistance in making the photographs.

Acquisition Editor: Susan Willig
Editorial/production supervision and
  interior design: Penelope Linskey
Cover design: Wanda Lubelska
Prepress Buyer: Ilene Levy
Manufacturing buyer: Ed O'Dougherty

© 1992 by Prentice Hall Career & Technology
Prentice-Hall, Inc.
A Pearson Education Company
Upper Saddle River, NJ 07458

9 10 11 CP 14 13 12

Printed in Canada

ISBN 0-13-249731-X

Prentice-Hall International (UK) Limited,London
Prentice-Hall of Australia Pty. Limited, Sydney
Prentice-Hall Canada Inc., Toronto
Prentice-Hall Hispanoamericana, S.A., Mexico
Prentice-Hall of India Private Limited, New Delhi
Prentice-Hall of Japan, Inc., Tokyo
Pearson Education Asia Pte. Ltd., Singapore
Editora Prentice-Hall do Brasil, Ltda., Rio de Janeiro

Contents

PREFACE   **xiii**

UNIT 1   **BASIC ELECTRICAL CIRCUIT COMPONENTS**   **1**

1-1   Fundamental Concepts   2

     1-1a   The basic electrical circuit, 2
     1-1b   Conductors, 4

         1-1b.1 Resistance of a conductor, 4

     1-1c   Types of wire, 6
     1-1d   Plugs, jacks, and cable connectors, 7
     1-1e   Insulators, 9
     1-1f   Types of switches, 9
     1-1g   Types of miniature and subminiature lamps, 11
     1-1h   The multimeter, 12
     1-1i   Circuit protectors, 14
     1-1j   Basic safety procedures, 15

         1-1j.1 Current hazards, 15
         1-1j.2 A common shock situation, 16
         1-1j.3 In case of electrical shock, 17
         1-1j.4 Other basic safety precautions, 18

1-2   Definition Exercises   19

1-3   Exercises and Problems   22

1-4   Experiments   24

     Experiment 1. Ohmmeter familiarization, 24
     Experiment 2. Testing a switch, 25
     Experiment 3. Testing a lamp, 25

1-5   Instant Review   26

1-6   Self-Checking Quiz   28

     Answers to Fill-in Questions and Self-Checking Quiz   29

UNIT 2   **PASSIVE ELECTRONIC DEVICES**   **31**

    2-1 Fundamental Concepts 32

        2-1a Resistors, 32

            2-1a.1 The carbon resistor, 32
            2-1a.2 Resistor color code, 32
            2-1a.3 Resistor sizes and power ratings, 34
            2-1a.4 Other types of resistors, 34
            2-1a.5 Variable resistors, 36
            2-1a.6 Potentiometer operation, 37
            2-1a.7 Nonlinear resistors, 38
            2-1a.8 Units of measurement, 39

        2-1b Capacitors, 42

            2-1b.1 Types of capacitors and their use, 42
            2-1b.2 Capacitor color code, 45

        2-1c Inductors, 45
        2-1d Transformers, 46
        2-1e Miscellaneous Devices, 46

            2-1e.1 The Relay, 46
            2-1e.2 Motors, 49
            2-1e.3 Audio Devices, 49
            2-1e.4 Circuit Indicators, 50
            2-1e.5 Other Devices, 50

    2-2 Definition Exercises 52

    2-3 Exercises and Problems 56

    2-4 Experiments 63

    Experiment 1. Testing resistors, 63
    Experiment 2. Testing capacitors, 65
    Experiment 3. Testing inductors and transformers, 67

    2-5 Instant Review 67

    2-6 Self-Checking Quiz 69

    Answers to Fill-in Questions and Self-Checking Quiz 71

UNIT 3   **ACTIVE ELECTRONIC DEVICES**   **73**

    3-1 Fundamental Concepts 74

        3-1a Diodes, 74

            3-1a.1 Normal diodes, 74
            3-1a.2 Light-emitting diode, 74
            3-1a.3 Other types of diodes, 75
            3-1a.4 Rectifier Modules, 76

        3-1b Three- and four-element semiconductor devices, 76

            3-1b.1 Bipolar transistors, 76
            3-1b.2 Field-effect transistors, 77
            3-1b.3 Thyristors, 77
            3-1b.4 Power devices and heat sinks, 78

        3-1c Integrated circuits, 80
        3-1d Optoelectronic devices, 82

            3-1d.1 Photodetectors, 82
            3-1d.2 Optoisolators, 83

        3-1e The schematic diagram, 83

    3-2 Definition Exercises 85

3-3 Exercises and Problems 88

3-4 Experiments 91

Experiment 1. Testing semiconductor diodes, 91
Experiment 2. Testing bipolar transistors, 92
Experiment 3. Testing JEETs, 93
Experiment 4. Testing a UJT, 94
Experiment 5. Testing an SCR with an Ohmmeter, 95
Experiment 6. Testing a TRIAC with an ohmmeter, 96
Experiment 7. Testing photodetectors, 97

3-5 Instant Review 99

3-6 Self-Checking Quiz 100

Answers to Fill-in Questions and Self-Checking Quiz 101

UNIT 4   **BASIC TOOLS AND THEIR USE**   **103**

4-1 Fundamental Concepts 104

    4-1a Basic hand tools, 104

        4-1a.1 *Tool description, 104*
        4-1a.2 *Using the hacksaw, 106*
        4-1a.3 *Using pliers, 107*
        4-1a.4 *Using wrenches, 107*
        4-1a.5 *Using screwdrivers, 109*
        4-1a.6 *Hex wrench, 109*
        4-1a.7 *Utility cutter, 110*
        4-1a.8 *Metal files, 111*
        4-1a.9 *Electrician's 6-in-1 tool, 111*

    4-1b Electronic hand tools, 112

        4-1b.1 *Wire stripper, 113*
        4-1b.2 *Diagonal cutters, 113*
        4-1b.3 *Long nose pliers, 114*

    4-1c Basic electric tools, 114

        4-1c.1 *The hand drill, 114*
        4-1c.2 *The jigsaw, 116*
        4-1c.3 *The drill press, 117*
        4-1c.4 *The bench grinder, 117*
        4-1c.5 *Sheet metal tools for chassis fabrication, 118*

    4-1d Tool maintenance, 118
    4-1e General safety precautions, 119

        4-1e.1 *Personal safety rules, 120*
        4-1e.2 *Hand tool safety rules, 120*
        4-1e.3 *Power tool safety rules, 120*
        4-1e.4 *General workshop or laboratory safety rules, 121*

4-2 Definition Exercises 121

4-3 Exercises and Problems 124

4-4 Experiments 127

Experiment 1. Using a hacksaw, 127
Experiment 2. Using a hand drill, 127
Experiment 3. Stripping insulation from wire, 128

4-5 Instant Review 129

4-6 Self-Checking Quiz 130

Answers to Fill-in Questions and Self-Checking Quiz 132

Contents

UNIT 5   **HARDWARE AND MECHANCIAL ASSEMBLY**   **133**

5-1   Fundamental Concepts   134

5-1a   General hardware used in electronics, 134
5-1b   Screw and bolt classification, 136

5-1b.1 Screw and bolt use or application, 136
5-1b.2 Type of head, 137
5-1b.3 Gage and size, 138
5-1b.4 Types of screw drives, 140
5-1b.5 Types of nuts, 141

5-1c   Standard mounting procedures, 141

5-1c.1 Mounting two flat pieces, 142
5-1c.2 Mounting a solder lug, 142
5-1c.3 Mounting a cable clamp, 143
5-1c.4 Mounting a grommet, 143
5-1c.5 Mounting spacers and stand-offs, 143
5-1c.6 Mounting a potentiometer and knob, 144
5-1c.7 Mounting a toggle switch, 145
5-1c.8 Mounting a slide switch, 146
5-1c.9 Mounting a snap-in switch, 146
5-1c.10 Mounting a fuse holder, 146
5-1c.11 Mounting LEDs, 146
5-1c.12 Mounting a strain relief, 148

5-2   Definition Exercises   149

5-3   Exercises and Problems   151

5-4   Experiments   153

Experiment 1. Standard parts mounting, 153
Experiment 2. Mounting a potentiometer, 154

5-5   Instant Review   155

5-6   Self-Checking Quiz   157

Answers to Fill-in Questions and Self-Checking Quiz   158

UNIT 6   **SOLDERING TECHNIQUES**   **159**

6-1   Fundamental Concepts   160

6-1a   Solder, 160

6-1a.1 Reasons for soldering, 160
6-1a.2 Composition of solder, 160
6-1a.3 Wetting action, 162
6-1a.4 Flux and acid, 162

6-1b   Soldering tools and methods, 163

6-1b.1 Basic Soldering Iron, 163
6-1b.2 Basic Soldering Tools, 163
6-1b.3 Proper soldering iron preparation, 164
6-1b.4 General soldering method, 165
6-1b.5 Types of soldering joints, 166
6-1b.6 Types of soldering tips, 167
6-1b.7 Temperature control, 168
6-1b.8 Tinning a wire, 169
6-1b.9 Splicing a wire, 169
6-1b.10 Removing solder, 171

6-1c   Soldering terminals, 173

6-1c.1 Turret terminal, 173
6-1c.2 Bifurcated terminal, 174
6-1c.3 Cup terminal, 175

6-1d   The soldering gun, 175

6-2   Definition Exercises   177

6-3   Exercises and Problems   180

6-4   Experiments   181

Experiment 1. Stripping and tinning a wire, 181
Experiment 2. Splicing wires, 182
Experiment 3. Soldering terminals, 183
Experiment 4. Removing solder from a solder joint, 184

6-5   Instant Review   185

6-6   Self-Checking Quiz   187

Answers to Fill-in Questions and Self-Checking Quiz   188

UNIT 7   **PRINTED CIRCUIT BOARD ASSEMBLY**   **189**

7-1   Fundamental Concepts   190

7-1a   Printed circuit board construction, 190

7-1a.1 Base or substrate material, 191
7-1a.2 Method of producing circuitry, 191
7-1a.3 Through-holes for components, 193
7-1a.4 Types of printed circuit boards, 193

7-1b   Mounting passive components, 195

7-1b.1 Forming component axial leads, 196
7-1b.2 Methods of lead termination to a PC board, 197
7-1b.3 Mounting the component, 198
7-1b.4 Solder procedures for component leads in PC boards, 199
7-1b.5 Mounting radial-lead components, 201

7-1c   Mounting active components, 202

7-1c.1 Mounting multi-lead components, 202
7-1c.2 Mounting DIP integrated circuits, 203

7-1d   Surface mount technology, 205

7-1d.1 Types of surface-mountable components, 205

7-1e   Soldering Component Sockets, 213
7-1f   Component replacement, 213

7-1f.1 Simple component replacement, 213
7-1f.2 Conformal coating removal, 214
7-1f.3 Removing solder from leads, 214
7-1f.4 Replacing components in a PC board, 215

7-2   Definition Exercises   216

7-3   Exercises and Problems   220

7-4   Experiments   223

Experiment 1. Soldering axial-lead components onto PC boards, 223
Experiment 2. Soldering ICs onto PC boards, 223
Experiment 3. Soldering surface mount components, 224

7-5   Instant Review   225

7-6   Self-Checking Quiz   227

Answers to Fill-in Questions and Self-Checking Quiz   228

UNIT 8   **WIRING PROCEDURES**   **229**

8-1   Fundamental Concepts   230

8-1a   Wire cables, 230

8-1a.1 Assembling wire cables, 230

                  8–1a.2 Using the nylon self-locking tie, 231
                  8–1a.3 Tying a cable with linen cord, 232

        8–1b The wire harness, 233
                  8–1b.1 Typical harness, 233

        8–1c Cable connectors, 234
                  8–1c.1 Solder connector, 234
                  8–1c.2 Snap-in pin connector, 235
                  8–1c.3 AMP cable connector with multiple parts, 235
                  8–1c.4 Flat (ribbon) cable connector, 236
                  8–1c.5 Shielded-cable connections, 237

        8–1d Wire wrapping techniques, 238
                  8–1d.1 Wire wrapping materials and tools, 238
                  8–1d.2 Wire wrapping procedures, 238

    8–2 Definition Exercises 239

    8–3 Exercises and Problems 241

    8–4 Experiments 243

    Experiment 1. Making a cable with sleeving, 243
    Experiment 2. Lacing with linen cord, 243
    Experiment 3. Assembling cable connectors, 244
    Experiment 4. Wire wrapping techniques, 245

    8–5 Instant Review 245

    8–6 Self-Checking Quiz 246

    Answers to Fill-in Questions and Self-Checking Quiz 247

UNIT 9   **REPAIRING ELECTRONIC DEVICES AND PC BOARDS**   **249**

    9–1 Fundamental Concepts 250

        9–1a The repair process, 250
                  9–1a.1 Types of repair, 250
                  9–1a.2 Elements of repair, 251
                  9–1a.3 Personal attributes for successful repair, 251
                  9–1a.4 Proper repair equipment, 252

        9–1b Repair station safety precautions, 253

        9–1c General repair work, 254

        9–1d Repairing damaged printed circuit boards, 255
                  9–1d.1 Repairing conductors, 256
                  9–1d.2 Repair of edge connectors, 256
                  9–1d.3 Repair of substrate, 257

    9–2 Definition Exercises 258

    9–3 Exercises and Problems 259

    9–4 Experiment 261

    Experiment 1. Repairing electronic equipment, 261

    9–5 Instant Review 261

    9–6 Self-Checking Quiz 262

    Answers to Fill-in Questions and Self-Checking Quiz 263

UNIT 10   **AUTOMATED ELECTRONIC ASSEMBLY**   **265**

    10–1 Assembly Line Procedures 266

        10–1a General PC board assembly line, 266
        10–1b Wave soldering, 267
        10–1c Automated assembly line, 268
                  10–1c.1 Insertion mount component machines, 269

10-1d Surface mount technology assembly, 272

    10-1d.1 PC board printing system, 272
    10-1d.2 SMC mounting machine, 274

10-1e Opportunitites for employment, 274

10-2 Definition Exercises   274

10-3 Exercises and Problems   275

10-4 Experiment   276

10-5 Instant Review   277

10-6 Self-Checking Quiz   278

Answers to Fill-in Questions and Self-Checking Quiz   278

APPENDIX A   **MISCELLANEOUS RESISTOR COLOR CODING**   **279**

APPENDIX B   **MISCELLANEOUS CAPACITOR COLOR CODING**   **281**

APPENDIX C   **INTERPRETING NUMBER/LETTER-CODED CAPACITOR VALUES**   **283**

INDEX   **285**

Preface

The electronics industry has changed considerably in the past decade. Many manufacturing techniques have changed and automation has taken over most of the jobs originally performed by workers. However, there is wide diversity among manufacturing facilities. Some plants may be totally automated, whereas others combine automatic machines and manually operated procedures in their production process.

Fabrication of electronic devices involves efforts of a technical team composed of engineers, electronic assemblers, and technicians. The original ideas for an electronic product are developed by the engineer, who may also plan how the device is to be fabricated. The electronics assembler manufacturers the product following the plans of the engineer. Testing the finished product to see that it meets specifications and functions properly is the job of the electronics technician. The technician is also responsible for the maintenance and repair of the equipment for its continued operation.

Jobs in electronics are suited to men and women of all nationalities, and many persons have achieved a high degree of success in all phases of the industry. Regardless of the new methods and techniques of manufacturing that appear nearly every day, a person desiring to enter the electronics industry must have the necessary basic understanding of electronic components and skills in the use of hand tools and soldering.

This book was written for those persons working with or planning to work with the production, repair, or maintenance of electrical and electronic equipment and devices. A list of such persons would include production assembly persons, component repair persons, technicians, field engineers, and systems engineers. Persons entering the electronics field as assemblers, testers, or technicians cannot merely go to work and pick up knowledge of the trade on the job; they must be trained. Even job-entry positions require that a person have some basic knowledge and skills in electronics. In addition, a person with knowledge and skills advances more rapidly in the electronics industry.

BOOK OBJECTIVES

When you have completed this book, you will be able to:

1. Explain a basic electrical circuit.
2. List safety practices associated with electrical/electronic devices and circuits.
3. Identify the schematic symbols of passive and active electronic components.
4. Recognize and use basic hand tools.
5. Prepare a soldering iron properly for soldering components.
6. Use correct soldering techniques for various soldering joints.
7. Define the process of creating a printed circuit board.
8. Assemble insertion-type components and surface mount components into printed circuit boards.
9. Produce cables and attach connectors to wires.
10. Repair basic component problems associated with electronic devices and printed circuit boards.
11. Show how assembly persons can upgrade themselves to meet the challenge of automation.
12. Explain the various terms used in fabricating and manufacturing electronic devices and equipment.

BOOK FORMAT

Each unit in this book follows a standard format consisting of six sections. In Section 1 the reader is introduced to the theory or information of basic topics. Section 2 consists of definition exercises that enable the reader to begin to understand and use the language of electronics assembly. Exercises and problems are given in Section 3 to develop in the reader a working knowledge of electronic devices and assembly procedures. These exercises include drawings, matching symbols, and various other methods to motivate and maintain the interest of the reader. In Section 4 we show how to perform simple tests on electronic devices, mechanical assembly, soldering procedures, and related methods of manufacturing. An instant review of basic terms is presented in Section 5 to reinforce the reader's knowledge. A self-checking quiz is given in Section 6 to provide the reader with instant feedback of what was learned from the unit. Answers are provided at the end of each unit.

FOR THE STUDENT

Manipulative skills such as mechanical assembly and the use of proper soldering procedures are essential for any electronics assembler and technician. However, the same person attempting this type of work must have knowledge of the terms used, be able to recognize schematic symbols, be aware of safety precautions for the protection of personnel, equipment, and tools, and be familiar with other manufacturing methods used in the industry.

This book is designed to be used in conjunction with a classroom setting, but is written as a self-study method. The material is presented in an easy-to-read, straightforward manner that will provide the readwer with a solid foundation for further study in electronics. The important thing to remember is to be serious about learning electronic assembly procedures and to spend time studying those topics that are most difficult for you.

FOR THE INSTRUCTOR

This book uses an individualized learning approach, which involves the reader in the activities of learning. This provides the instructor with student-centered instruction that does not require preparation. In a lock-step type of course, students still have some capability to learn at their own rate with the use of this book. A student is able to perform other sections of a unit while waiting for personal instruction from the instructor. By placing more responsibility on students for their learning, schools are able to graduate better-prepared persons to enter the work force.

I wish you good luck and success in the electronics industry.

Fredrick W. Hughes

Unit 1

Basic Electrical Circuit Components

INTRODUCTION   Anyone working with the construction and assembly of electrical/electronic devices must be able to identify basic components and have a knowledge of basic circuit operation. This not only enhances a person's skills, but provides a foundation for understanding safety when working with tools and the prevention of electrical shock.

UNIT OBJECTIVES   Upon completion of this unit, you will be able to:

1. Define the terms *voltage, resistance, current, volt, ohm,* and *ampere.*
2. Describe a basic electrical circuit.
3. Identify types of wire, switches, lamps, and safety devices.
4. State what constitutes a good electrical conductor.
5. List materials that make good conductors.
6. State what constitutes a good electrical insulator.
7. List materials that make good insulators.
8. Define basic wire and cable connectors.
9. Describe a multimeter.
10. Use an ohmmeter to test a lamp and a switch.
11. Define *open circuit, closed circuit,* and *shorted circuit.*
12. Explain how an electrical shock occurs.
13. Show what is meant by *grounding* a circuit.
14. List good safety practices when working with electrical/electronic equipment and devices.

SECTION 1–1
FUNDAMENTAL CONCEPTS

1–1a   THE BASIC ELECTRICAL CIRCUIT

The basic electrical circuit is easily illustrated with the common battery-operated flashlight, shown in Figure 1–1. To produce light you must have a power source, such as batteries; a lamp; electrical conductors to connect the lamp to the batteries; and a switch to provide on/off control. Usually, these components are placed in an insulating plastic case with a reflector, which increases the light's brightness. The bottom part of the switch has a metal portion that contacts the conductor with a forward movement and allows electrical current to flow through the lamp, producing light.

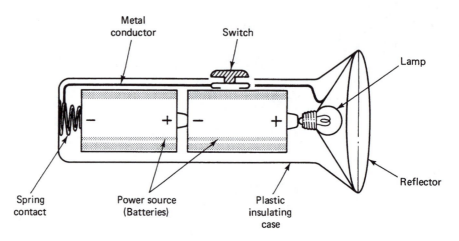

Figure 1–1   Simple flashlight (cross-sectional view).

An *electrical circuit* can be defined as an electrical path between two or more devices for the purpose of carrying electrical current to produce a useful function. There are three important terms associated with an electrical circuit:

1. *Voltage:* the force that causes electrical current to flow in a circuit. Voltage is the difference between two charges, such as a battery. The negative end of the battery has a large number of negatively charged particles, referred to as *electrons*. The positive end of the battery has many fewer electrons and, therefore, is referred to as a *positive charge*. When a conductor is placed across the terminals of a battery, the electrons will flow from the negative terminal through the conductor to the positive terminal of the battery. The *volt* (V), the unit of voltage, helps to identify various forms of voltage, such as a 6-, 9-, or 12-V battery. A battery operates on dc (direct current) voltage since the electrons always flow in one direction, from negative to positive. Typical house voltage changes polarity at a specified frequency of 60 cycles per second and the electrons flow first in one direction, then in the other. This type of voltage is referred to as ac (alternating current) voltage.

2. *Current:* the flow of electrons through the circuit caused by the force of voltage. The *ampere* (A), the unit of current, aids in showing amounts of current, such as 20 A, 10 A, or very small amounts such as 0.1 A [100 milliamperes (mA)].

3. *Resistance:* the opposition to current flow of the components and conductors in a circuit. The *ohm* [Ω (Greek capital letter omega)] the unit of resistance, helps to identify various values, such as 10 Ω, 100 Ω, 1000 Ω, and 1,000,000 Ω.

A flashlight can be represented by a basic electrical circuit, as shown in Figure 1–2. The power source is a 1.5-V battery, the load, which is an electrical device that performs a useful function; in this case, a lamp and the conductors are connecting wires. A switch is also used to control the flow of electrical current in the circuit.

In Figure 1–2 a, when the switch lever is placed to the left, the switch is open and no current flows in the circuit. When the switch lever is placed to the right, the switch is closed and electrons leave the negative terminal of the battery and flow through the switch, then through the lamp to the positive terminal of the battery. The electrons flowing through the resistance of the lamp create heat and produce light—hence, a typical flashlight circuit.

Figure 1–2b and c show a *schematic diagram* (electrical drawing). Notice the symbols for battery, switch, and lamp. As a technician, you should be able to draw these symbols from memory. In Figure 1–2b the switch is open and the resistance of the circuit from the viewpoint of the battery is infinite. Therefore, no current flows in the circuit and the lamp is off or not glowing. This condition is referred to as an *open circuit*. In Figure 1–2c the switch is closed and current flows through the circuit, as indicated by the arrows, which in turn causes the lamp to glow. The battery now "sees" a lower resistance, which depends on the resistance of the lamp. The closed switch is assumed to have no or zero resistance. This condition is referred to as a *closed circuit*.

The circuit in Figure 1–2 is a *series circuit,* because electrons leaving the negative terminal of the battery flow through the switch and then through the lamp to reach the positive terminal of the battery. Since the

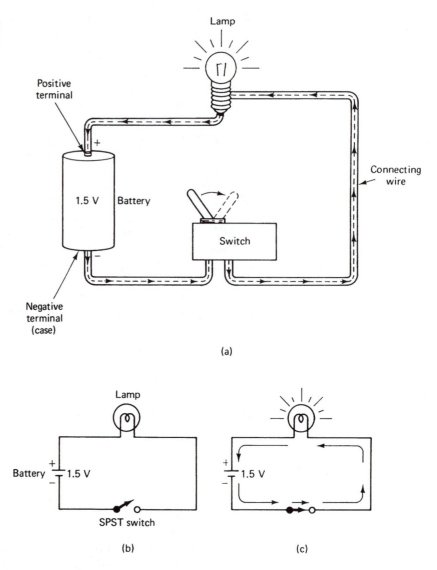

Figure 1-2 Basic electrical circuit: (a) pictorial diagram; (b) schematic diagram, open circuit; (c) schematic diagram, closed circuit.

same current flows through the switch and then the lamp, it is said that the switch and lamp are in series.

1-b CONDUCTORS

An *electrical conductor* is a material that has a large number of electrons capable of flowing through the material. These electrons are often referred to as *free electrons*. Most good conductors are metallic in substance. Silver is the best known practical electrical conductor. Table 1-1 shows a comparison of some conductive materials with a diameter of 0.001 in. (1 mil) and a length of 1 ft at a constant temperature of 20°C.

TABLE 1-1

Comparison of Conductor Materials

Material	Ohms/mil, foot	Relative Conductivity
Silver	9.8	1 (high)
Copper	10.4	2
Gold	14.7	3
Aluminum	17.0	4
Tungsten	33.2	5
Nichrome	660.0	6 (low)

Figure 1–3 Resistance of a conductor.

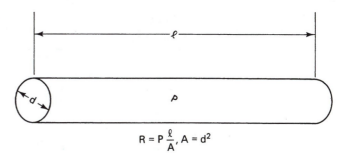

$$R = P\frac{\ell}{A}, A = d^2$$

Some metals are very expensive and cannot be found in large quantities. Copper, second in the list of conductivity values, is readily available and used primarily in the manufacture of electrical wire.

1–1b.1 Resistance of a Conductor

The principal factors determining the resistance of a conductor are the type of material it is made of, represented by the Greek lowercase letter rho (ϱ), its cross-sectional area, and its length. Figure 1–3 shows the relationship of these factors. The resistance of a conductor varies in direct proportion with its length and in inverse proportion with its cross-sectional area. Stated in a formula, we have

$$R = \varrho\frac{l}{A}$$

where P = resistance of the material
ℓ = length of the conductor (feet)
A = cross-sectional area (circular mils),
which is equal to the area of a circle
0.001 in. (1 mil) in diameter (also,
$A = d^2$)

In other words, with two wires of the same thickness, the longer wire will have more resistance than the shorter one. If there are two wires of the same length but of different thickness, the thick wire will have less resistance than the thin wire. As an example, a large garden hose with a larger diameter will allow more water to flow than will a smaller garden hose with a smaller diameter.

A comparison of the cross-sectional area of several commonly used wires is shown in Figure 1–4. Notice the difference in size between wires. The larger the number, the smaller the wire and the less current-carrying capacity.

"AWG" stands for *American Wire Gage,* a standard used for wire sizes. Table 1–2 compares these wires with diameter, resistance, and safe current-carrying capacity.

Figure 1–4 Approximate cross-sectional area comparison of several wire sizes.

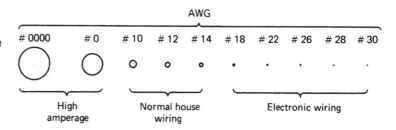

TABLE 1–2

Comparison of Wire Sizes

Gage	Diameter (mils)	Resistance/100 ft	Current Capacity (A)
0000	460.2	0.0049	225
0	325.0	0.0098	125
10	101.9	0.099	25
12	80.8	0.159	20
14	64.0	0.252	15
18	40.3	0.639	3
22	25.4	1.614	< 1
26	15.9	4.08	< 1
28	12.6	6.49	< 1
30	10.0	10.30	< 1

1–1c TYPES OF WIRE

Various types of wire are used in electrical and electronic circuits and devices. Figure 1–5 shows a few of these types and the following is a brief description of their use.

Solid wire is easy to handle but cannot withstand much flexing. Bare solid wire is used for buses, a common connection line for components

Figure 1–5 Types of wire: (a) solid strand with/without insulation; (b) insulated solid strand; (c) insulated stranded wire; (d) two-conductor wire; (e) shielded wire; (f) coaxial cable; (g) interconnecting (three-wire) cable; (h) special-purpose interconnecting cable; (i) flat "ribbon" cable.

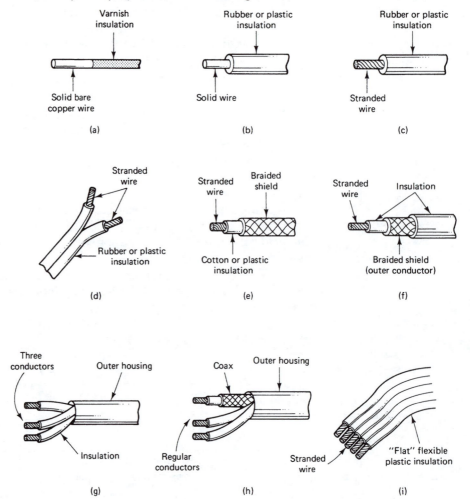

and other wires. Solid wire with varnish or paint insulation (a) is used for windings of coils, transformers, generators, and motors. Insulated (rubber or plastic) solid wire (b) has advantages in some circuits. Small (AWG No. 30) insulated solid wire is best to use for integrated circuits (ICs) and connections that are very close.

Flexible wire consists of several strands of twisted wire covered by insulation (rubber, paper, plastic, Teflon, etc.) and is referred to as *stranded wire* (c). Stranded wire is used extensively in electronic circuits. Two stranded wires may be joined together to form a *two-conductor wire* (d), such as that used in electrical appliances.

Shielded wire (e) consists of a stranded inner conductor covered with insulation and then a braid of woven segments of stranded wire. This shield is used as a conductor and also to block electrical interference from the inner conductor. Some shields are made of tin foil with an unshielded wire that is used for connecting purposes. *Coaxial cable* (f) is similar to shielded wire, except that an insulated covering is used around the shield.

Several wires can be put into a larger plastic tube interconnecting cable. Some of these cables may contain shielded or coaxial cable and other special arrangements (g, h). The newer type of flat "ribbon" cable has stranded wire sizes of AWG Nos. 22 to 30, which are connected side by side, with up to 64 conductors in a single cable. Ribbon cables are used primarily in computer installations.

1–1d PLUGS, JACKS, AND CABLE CONNECTORS

Plugs, jacks, and cable connectors facilitate wire connections between various electronic circuits and equipment. *Plugs* are usually attached to the wire, while their receptable or *jack* is rigidly mounted to a chassis or some equipment. A plug fits into a jack. Plugs are shown in Figure 1–6.

Phone plugs (a) are two or more wire connectors that are used in telephone central offices, radio–television broadcasting, and professional entertainment (musical instrument amplifiers and microphones). The smaller *phono plug* (b) is a two-wire connector used in audio systems (i.e., from turntable to amplifier, from tape deck to amplifier, etc.). Two-wire conductors and shielded and coaxial cables are used with phone and phono plugs. A *pin plug* (c) is a single-wire connector used where circuit changes are sometimes required. The *banana plug* (d) has the same function as the pin plug, except that it is larger and the spring connectors on the tip can be adjusted to form a better electrical connection.

Figure 1–6 Plugs: (a) phone; (b) phono; (c) pin; (d) banana.

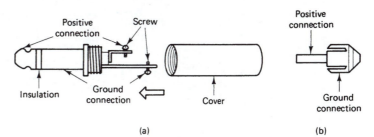

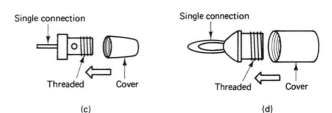

Figure 1–7 Jacks:
(a) phone; (b) phono;
(c) pin; (d) banana.

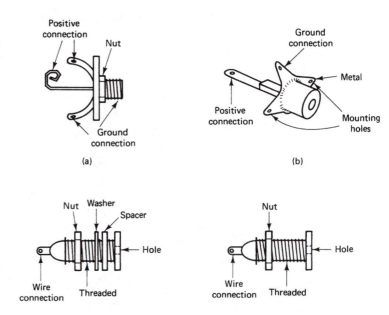

The *phone jack* is a connector of two or more wires that matches the phone plug. It may also have contacts that open or close circuits when the plug is inserted. A *phono jack* is a two-wire connector that accepts the phono plug. The pin and banana jacks are single-wire connectors that match their respective plug types and usually are mounted through a panel or chassis. Jacks are shown in Figure 1–7.

Coaxial-cable connectors are used on high-frequency equipment and test equipment. There are both *screw-type* and *bayonet-type connectors,* which provide good electrical connections. *Pin-contact connectors* require a *male* half and a *female* half. A keyway and key are used on these connectors to align the pins and socket holes properly. *Rack-and-panel connectors* are used with subassemblies or subchassis that fit into a large mounting rack. This connector serves as a quick-disconnect device for

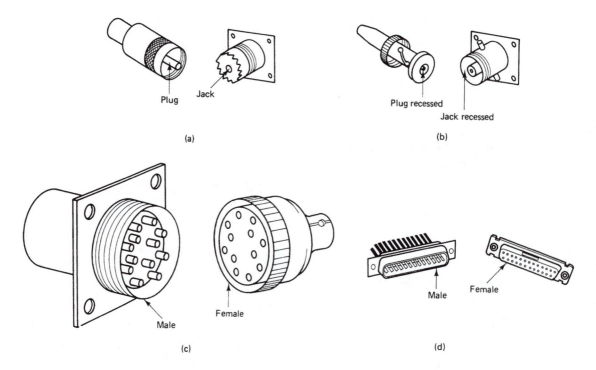

Figure 1–8 Cable connectors: (a) coaxial connector (screw type); (b) coaxial connector (bayonet type); (c) pin-contact-type connector; (d) rack-and-panel connector.

Chap. 1 / Basic Electrical Circuit Components

rapid test or replacement of subassemblies. Cable connectors are shown in Figure 1–8.

1–1e INSULATORS

An *electrical insulator* is a material that has very few free electrons and opposes the current flow through the material. Insulators and insulating material are used with wire and components to protect people from getting shocked or electrocuted and to prevent short circuits from destroying circuits and causing fires. Wires are insulated by means of rubber, plastic, or Teflon covering. Rubber gloves are worn by utility employees working on high-voltage power lines. The power lines themselves are insulated from the power poles by glass and/or porcelain insulators. Printed circuit boards are made from plastic, Bakelite, or fiberglass to insulate electronic components from each other.

Even though insulators oppose the flow of electron current, sufficient voltage can be placed on a material to cause current flow. The voltage is extremely high and a point is reached where electrons will begin to flow. The insulating material is usually destroyed and in many cases will burn, which creates a carbon residue and allows more electrons to flow. The high-voltage point at which electrons begin to flow in an insulator is called the *breakdown voltage*. Table 1–3 shows a comparison of different types of insulating materials.

TABLE 1–3

Comparison of Insulating Materials

Material	Resistance (Ω/cm)	Relative Insulating Ability	Breakdown Voltage (kV/cm)
Glass	10^{14}	1 (high)	1200
Teflon	10^{14}	2	600
Rubber	10^{13}	3	270
Bakelite	10^{12}	4	150
Waxed paper	10^{9}	5	500
Air (dry)	—	6 (low)	30

1–1f TYPES OF SWITCHES

A *switch* is used to open or close an electrical circuit. The switch can transfer electrical current from one circuit to another; in other words, it can transfer current flow from one wire to another wire. There are many types of switches for various applications, but the function remains the same: to open or close a circuit. Figure 1–9 shows the schematic symbols for various basic types of switches.

The simplest switch is a single-pole single-throw, SPST (a), which closes a circuit from a common contact, C, to a normally open, NO, contact. This switch will have two terminals for connecting the wires. A single-pole double-throw switch, SPDT (b), transfers the current from one normally closed, NC, contact to a normally open, NO, contact. This switch will have three termianls for connecting wires. A double-pole single-throw switch, DPST (c), operates similarly to a SPST switch, except that the two C contacts operate simultaneously to open or close two separate circuits. This action, where two or more contacts are moved at the same time, is referred to as being *ganged together*. The DPST switch will have four terminals for connecting wires. The double-pole double-throw switch, DPDT (d), has the C contacts ganged together and operates two NC and NO contacts simultaneously. This switch has six terminals for connecting wires. A normally open pushbutton switch (e) closes a circuit when

Figure 1–9 Switch schematic symbols:
(a) single-pole single-throw;
(b) single-pole double-throw; (c) double-pole single-throw; (d) double-pole double-throw;
(e) normally open pushbutton; (f) normally closed pushbutton;
(g) multiposition; (h) rotary.

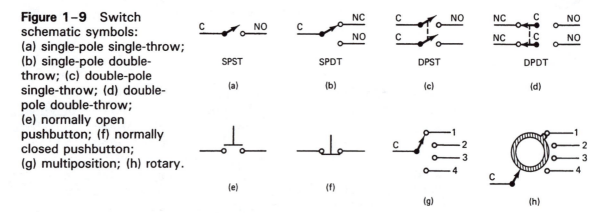

the button is depressed, whereas the normally closed switch (f) opens the circuit. The switch may keep the circuit latched in the desired condition when the button is released, or it may be spring loaded and return the contacts to the original condition. A pushbutton switch may operate several NO or NC contacts or a combination of both simultaneously. In a multiposition switch (g and h), also referred to as a *rotary switch,* the

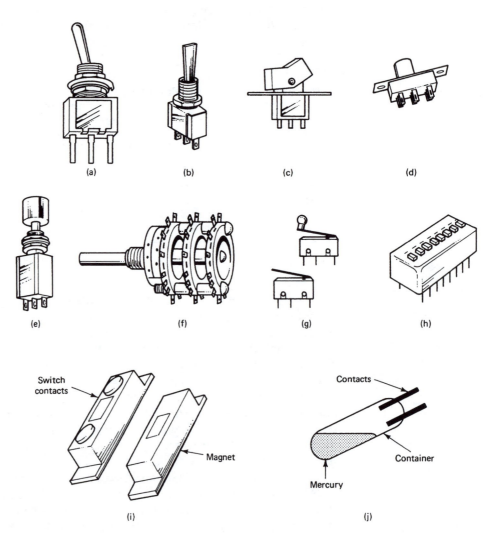

Figure 1–10 Types of switches: (a) toggle; (b) paddle; (c) rocker; (d) slide; (e) pushbutton; (f) multiwafer rotary; (g) subminiature (micro); (h) PC board eight-position; (i) proximity (magnetic); (j) mercury.

C contact can be set to many position contacts on a wafer-type form. There can be many wafers in a single switch, providing many electrical circuit transfers.

Switches are also classified as to their physical type, as shown in Figure 1–10. The distinguishing feature among switches is the type of C contact actuator used. A *toggle switch* (a) has a small round lever that moves back and forth. A *paddle switch* (b) is like the toggle except that the lever is wider. A *rocker switch* (c) rocks back and forth. A *slide switch* (d) moves laterally from one side to the other. A *pushbutton swith* (e) moves down and up (in or out). A *rotary switch* (f) moves in a circular motion. *Subminiature* (or *micro*) *switches* (g) are tiny switches that operate with a small plunger or actuator arm, which may have a small roller. *Switch units used on PC boards* (h) are about the size of an IC and fit into a dual-in-line package (DIP) mounting arrangement. They usually have eight separate slide switches. *Proximity switches* (i) have thin metal reed contacts, which operate when an external magnet is brought close to a contact. This type of switch is often used in intrusion (burglar) alarm systems. A *mercury switch* (j) is a small container (often, glass) with a portion of a liquid mercury sealed inside. The contacts at one end protrude through the glass for external connections. When the container is tilted, the mercury flows around the contacts, which completes the electrical circuit. This type of switch is used to close or open circuits resulting from physical movement of mechanical parts, such as an interlock switch on a lid or cover leading to a dangerous piece of equipment.

1–1g TYPES OF MINIATURE AND SUBMINIATURE LAMPS

Miniature and subminiature electrical lamps are used primarily in flashlights and as pilot (on/off indicator) lights for electronic equipment. Figure 1–11 shows various types of lamps and lamp sockets. Incandescent lamps (a–d) have very thin filament wires that are heated when current flows through them. The wires, in turn, glow and produce light. There are two major considerations in the use of these lamps: (1) that the rated voltage of the lamp not be exceeded, and (2) that when replacing a specific lamp with a substitute, to be aware of a possible high current draw, which could activate the fuse in a piece of equipment. It is wise to know the voltage and current ratings of any lamps being used in circuits.

There are three standard bases for incadescent lamps: *screw* (b), *bayonet* (c), and *flange* (d). Each type of base has to match a corresponding type of socket for proper use.

The *neon lamp* (e) is a nonheated type of device, with two electrodes, that contains an inert gas. At a given voltage, the gas ionizes and current flows, causing the electrodes to glow. If used with dc voltage, one electrode glows, whereas both electrodes glow with ac use. This lamp requires a series resistor to limit the current through it.

The *light-emitting diode,* LED (f), is rugged, requires little power, come in various colors (red, yellow, green), and is relatively inexpensive. In operation, the LED requires a series current-limiting resistor. Its main disadvantage is that it may not produce enough light for some specific applications. Groups of these LEDs, in a seven-segment form, are found in digital clocks, microwave ovens, radios, television sets, and various other appliances and equipment.

Lamp sockets and indicators (g–j) are available with different-colored lenses and various package styles. Some indicators come complete with lamp, limiting resistor (if needed), and even a switch.

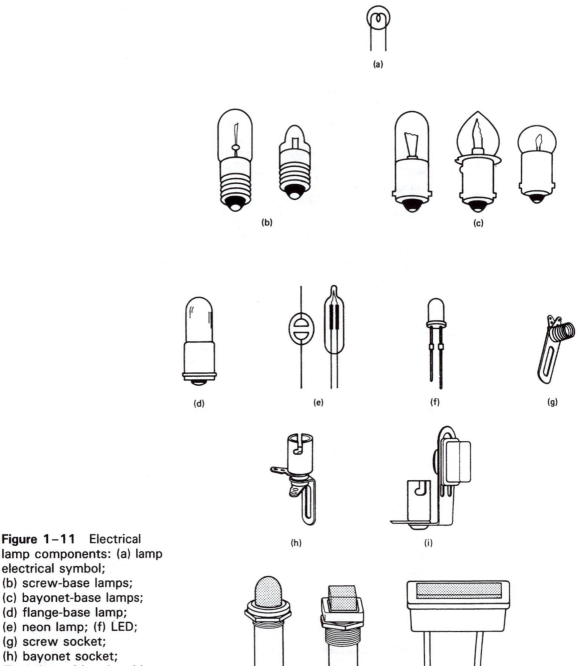

Figure 1–11 Electrical lamp components: (a) lamp electrical symbol;
(b) screw-base lamps;
(c) bayonet-base lamps;
(d) flange-base lamp;
(e) neon lamp; (f) LED;
(g) screw socket;
(h) bayonet socket;
(i) socket with colored lens;
(j) through-panel pilot light indicators.

1–1h THE MULTIMETER

Some means of testing inoperative circuits must be used in order to save time, if not money. For instance, with the basic electrical circuit of Figure 1–2b, when the switch is closed, the lamp does not glow. The question arises: Is the lamp burned out, the switch faulty, or the battery run down? New components can be purchased for each of the three and then each one substituted (replaced) in the circuit until the faulty component is located. If only one component is bad, extra money will not have to be spent replacing the other two new components.

Meters are used to check the operation of circuit components and actual circuit performance. Three meters are used to check the total performance of a circuit. An *ohmmeter* is used to measure the resistance

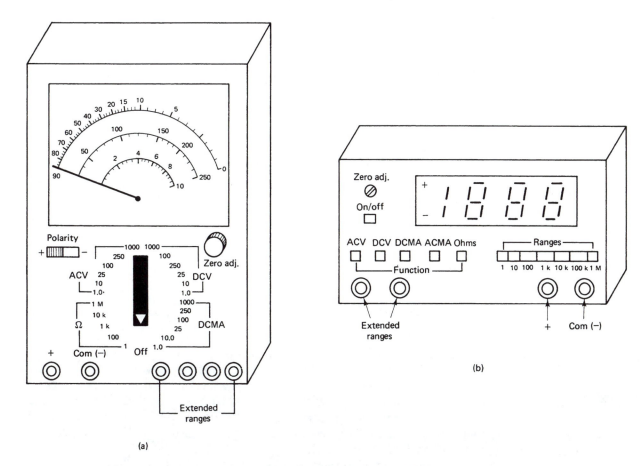

Figure 1-12 Multimeters: (a) analog VOM; (b) digital VOM (DVM).

of the lamp and operation of the switch. A *voltmeter* is used to measure the voltage of the battery. An *ammeter* is used to measure the current flowing in the circuit. These three meters are usually built into a single unit called a *multimeter*, shown in the generalized drawings of Figure 1–12.

An analog multimeter (often referred to as a *volt-ohm-ammeter* or VOM) has a set of scales on which a needle or pointer indicates a quantity being measured (Figure 1–12a). The digital VOM or digital voltmeter (DVM) uses seven-segment numerical readout indicators (Figure 1–12b). Each of these meters has a function switch or switches for selecting the measurement of resistance, voltage, or current. Generally, resistance is indicated by ohms or its symbol, Ω. Direct-current voltage, such as for measuring a battery, is labeled DCV. Direct current is given as DCMA, with the "MA" indicating milliamperes. Also included is a section for measuring effective alternating current voltage, given as ACV. Usually, an analog VOM does not measure alternating current, but most DVMs can and may be indicated as ACMA. The function switch may include ranges (specific limits to be measured) or have separate range switches.

There will be jacks for connecting test leads. The two most used jacks are the + jack and the common or − jack. Other jacks are used for extended ranges. A zero adjust control is used to adjust the meter for a reading of zero when resistance is being measured. This control is used to compensate for weakened battery voltage or internal circuitry required by the meter in the resistance measuring function.

A polarity switch may be included with the analog VOM, which simply reverses, electrically, the test lead jacks. If the leads are connected incorrectly in a circuit, the pointer will swing off the meter scale the wrong way. By moving this switch to the other position, the pointer will move correctly up the scale. The DVM usually has a + or − indicator that lights up to show proper polarity of the test leads. Following experiments will show you how to read an analog VOM correctly.

1–1i CIRCUIT PROTECTORS

Most electrical/electronic equipment, and even the electrical system in your house, has circuit protectors in the case of an overload. The most common circuit protector is the fuse, and the schematic symbol is shown in the circuit of Figure 1–13.

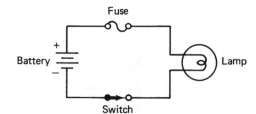

Figure 1–13 Circuit with fuse.

A fuse is normally found after the on/off switch mounted in a special holder or bracket. It is placed in series with the power source and the load or device. The fuse is made of a single strand of wire that is designed to burn open when its specific current rating is exceeded. The excessive current flowing through the fuse develops sufficient heat that the wire melts, which creates an open circuit. With this situation, the power source is removed from the circuit.

In the design of equipment, the fuse current rating selected is usually 10 to 50% greater than the total circuit current. The fuse will "operate" or burn open if one of two conditions occurs. If the power source voltage

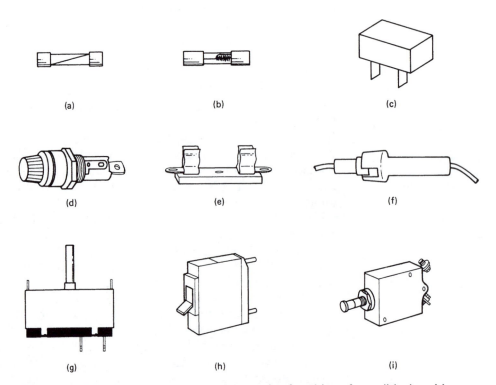

Figure 1–14 Circuit protectors: (a) regular fast-blow fuse; (b) slow-blow fuse; (c) bel fuse; (d) through-chassis fuse holder; (e) top-mount fuse holder; (f) in-line fuse holder (car radio); (g) television circuit breaker; (h) toggle-switch circuit breaker; (i) pushbutton circuit breaker.

increases, often referred to as *voltage surge*, the current in the circuit will increase. Voltage surges are usually of short duration and do not cause an excessive amount of current flow to damage some circuits. In this case a *slow-blow fuse* is used. It will continue to carry the excessive amount of current for a little longer, so that the voltage will probably return to normal before it burns open. In the other condition, the resistance of the load decreases or becomes shorted, which causes the circuit current to increase.

Figure 1–14 shows some types of circuit protectors. A circuit breaker is a thermal device that operates a switch when an excessive current develops enough heat. The circuit breaker normally can be reset after the problem in the circuit is corrected and the temperature decreases.

1–1j BASIC SAFETY PROCEDURES

The following information is very important to your safety and well-being. Read it carefully and study it sufficiently to understand the reason that people experience electrical shock.

1–1j.1 Current Hazards

It takes a very small amount of current passing through the human body from an electrical shock to injure a person severely or fatally. The 60-hertz (Hz) current values affecting the human body are as follows:

Current Value	Effect
1mA (0.001 A)	Tingling or mild sensation
10 mA (0.01A)	A shock of sufficient intensity to cause involuntary control of muscles, so that a person cannot let go of an electrical conductor
100 mA (0.1 A)	A shock of this type lasting for 1 second is sufficient to cause a crippling effect or even death
Over 100 mA	An extremely severe shock that may cause ventricular fibrillation, where a change in the rhythm of the heartbeat causes death almost instantaneously

The resistance of the human body varies from about 500,000 Ω when dry to about 300 Ω when wet (including the effects of perspiration). In this case, voltages as low as 30 V can cause sufficient current to be fatal (I = voltage/wet resistance = 30 V/300 Ω = 100 mA).

Even though the actual voltage of a circuit being worked on is low enough not to present a very hazardous situation, the equipment being used to power and test the circuit (i.e., power supply, signal generator, meters, oscilloscopes) is usually operated on 120 V ac. This equipment should have (three-wire) polarized line cords that are not cracked or brittle. An even better safety precaution is to have the equipment operate from an isolation transformer, which is usually connected to a workbench. To minimize the chance of getting shocked, a person should use only one hand while making voltage measurements, keeping the other hand at the side of the body, in the lap, or behind the body. Do not defeat the safety feature (fuse, circuit breaker, interlock switch) of any electrical device by shorting across it or by using a higher amperage rating than that specified by the manufacturer. These safety devices are intended to protect both the user and the equipment.

1–1j.2 A Common Shock Situation

Electrical shock occurs when electrical current passes through the human body. *Electrocution* is the result of death from an electrical shock. The common ac voltage in your home usually has three wires, as shown in Figure 1–15. The right-hand (R) opening to the electrical outlet is connected to the "hot side" of the power source coming into the house. The left-hand (L) opening of the outlet is connected to the other side of the power source and in most modern structures, also to earth ground. The third opening (GND) is connected directly to earth ground. Earth ground is provided by a connection to the metal water pipes or a metallic rod 8 ft long driven into the ground outside your house.

If you are standing on the ground and touch the L side of the electrical outlet, nothing will happen because you are at ground potential and there is no voltage potential to cause current flow. You are electrically at the same point. If you are standing on an insulated floor or special

Figure 1–15 Common shock situation: (a) insulated floor prevents current flow; (b) current flows through body to ground.

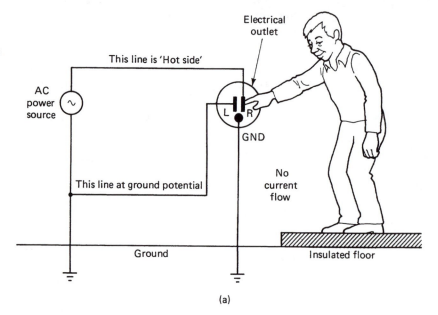

(a)

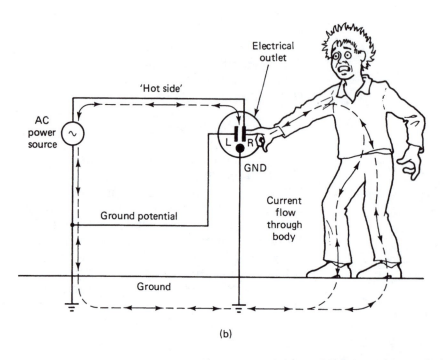

(b)

insulated surface and touch the R ("hot side") of the electrical outlet, nothing will happen because the insulated surface will not allow current to pass. It acts as an open circuit and there is no current flow. Your body is at the hot-side voltage potential, but because there is no current flow, you feel nothing (Figure 1–15a). The same principle is the reason why birds sitting on high-voltage electrical wires are not electrocuted. However, if you are standing on ground and touch the R side of the electrical outlet, your body will complete the path for current to flow back to the power source (Figure 1–15b). Your body acts as a switch that closes the circuit. In other words, you have shorted out the power source. You may be able to free yourself from the power source and only feel somewhat shaken; however, your muscles could contract and you might be caught by the force of the power source and be unable to let go, which could result in severe shock or even death.

Caution! If the electrical outlet is mounted upside down with the ground jack on top, the left or L opening is "hot." The best practice is to observe the safety rules and try not to come in contact with any electrical potential.

The earth ground wire on line cords used for electrical apparatus is to provide safety and reduce the chance of electrical shock to the person using the device. This is particularly important for electrical hand tools when working on cement floors or outside on the ground.

Do not fear electricity—just be very careful when working around it. Only a very small percentage of people of the vast number who work with or use electrical devices are shocked or electrocuted. **Be careful and observe safety procedures!**

1–1j.3 In Case of Electrical Shock

When a person comes in contact with an electrical circuit of sufficient voltage to cause shock, certain steps should be taken as outlined in the following procedure:

1. Quickly remove the victim from the source of electricity by means of a switch or circuit breaker, by pulling the cord, or by cutting the wires with a well-insulated tool.
2. It may be faster to separate the victim from the electrical circuit by using a dry stick, rope, leather belt, coat, blanket, or any other nonconducting material.
    Caution! Do not touch the victim or the electrical circuit unless the power is off.
3. Call for assistance, since other persons may be more knowledgeable in treating the victim or can call for professional medical help while first aid is being given.
4. Check the victim's breathing and heartbeat.
5. If breathing has stopped but the victim's pulse is detectable, give mouth-to-mouth resuscitation until medical help arrives.
6. If the heartbeat has stopped, use cardiopulmonary resuscitation, *but only if you are trained in the proper technique.*
7. If both breathing and heartbeat have stopped, alternate between mouth-to-mouth resuscitation and cardiopulmonary resuscitation (*but only if you are trained*).
8. Use blankets or coats to keep the victim warm and raise the legs slightly above head level to prevent shock.
9. If the victim has burns, cover your mouth and nostrils with gauze or a clean handkerchief to avoid breathing germs on the victim and

then wrap the burned areas of the victim firmly with sterile gauze or a clean cloth.

10. *In any case, do not just stand there*—do something within your ability to give the victim some first aid.

1–1j.4   Other Basic Safety Precautions

When too many electrical devices are plugged into an electrical outlet that has an adapter, as shown in Figure 1–16, too much current can be taken from the outlet, causing it to overheat. Each device that plugs into the electrical outlet represents a *load*. Too many loads cause an *overload* condition, where excessive current flows.

The line cord attached to an electrical device is susceptible to damage. As shown in Figure 1–17, it should never be detached by pulling the cord, but only by pulling the plug end. Pulling on the cord can result in a stress that breaks the insulated covering around the wires in the cord. A short circuit could result, which might start a fire.

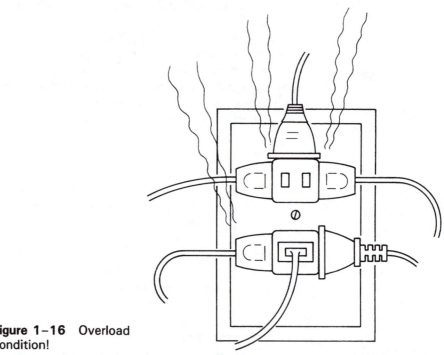

Figure 1–16   Overload condition!

Figure 1–17   A good way to break it!

Frayed insulation on wires or line cords (Figure 1–18) should be repaired before the cord is plugged into an electrical outlet, to prevent accidental shock or a short circuit that might result in a fire.

Frayed
insulation

Figure 1–18 Repair before use!

Water is a good conductor of electrical current. Therefore, remember not to work with electrical devices or apparatus when you are wet or damp. Be particularly careful about standing on wet surfaces, as shown in Figure 1–19.

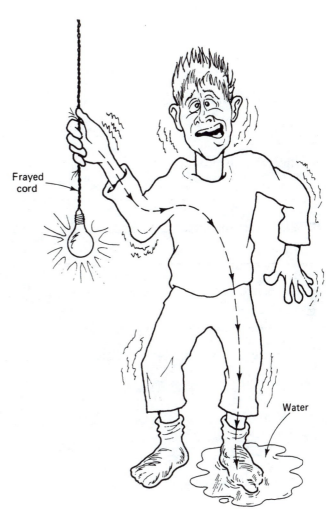

Frayed
cord

Water

Figure 1–19 Don't let this be you!

≡ **SECTION 1–2**
DEFINITION EXERCISES

Write a brief description of each of the following terms.

1. Voltage _____

_____

2. Volt _____

_____

3. Resistance _____

_____

4. Ohm _____

_____

5. Current_____

_____

6. Ampere _____

_____

7. Electrical circuit _____

_____

8. Electrical conductor _____

_____

9. Electrical insulator _____

_____

10. Electrical plug _____

_____

11. Electrical jack _____

_____

12. Schematic symbol _____

_____

13. Electrical switch _____

_____

14. SPST _____

_____

15. SPDT _____

_____

16. DPST _____

_____

17. DPDT _____

_____

18. Ganged together _____

_____

19. Open Circuit _____

_____

20. Closed circuit _____

_____

21. Short circuit _____

_____

22. Incandescent lamp _____

_____

23. Neon lamp _____

_____

24. LED _____

_____

25. Ohmmeter _____

_____

26. Voltmeter _____

_____

27. Ammeter _____

_____

28. Multimeter _____

_____

29. Fuse _____

_____

30. Slow-blow fuse _____

_____

31. Circuit breaker _____

_____

32. Electrical shock _____

_____

33. Electrocution _____

_____

34. Load _____

_____

35. Overload _____

_____

36. Earth ground _____

_____

37. Voltage surge _____

_____

38. Frayed wire _____

_____

≡≡≡ SECTION 1–3
EXERCISES AND PROBLEMS

Complete this section before beginning the next section.

1. Draw the electrical symbol for the following components.

    a. battery   **b.** lamp   **c.** SPST switch   **d.** Fuse

2. List the names of six materials that are good conductors.

    a.   **b.**

    c.   **d.**

    e.   **f.**

3. Circle the most correct answer for each question.

   a. When the length of a wire increases, its resistance: increases, decreases.

   b. When the cross-sectional area of a wire decreases its resistance: increases, decreases.

   c. An AWG No. 22 wire can carry more current than a No. 18 wire. True, False.

   d. Copper is a better conductor than aluminum. True, False.

4. Match the type of wire in column A with its appropriate use in column B.

   Column A

   ____ **a.** Solid wire

   ____ **b.** Stranded wire

   ____ **c.** Two-conductor

           stranded wire

   ____ **d.** Shielded wire

   ____ **e.** Ribbon cable

   Column B

   1. table lamp

   2. electronic circuits

   3. motor

   4. computer

   5. blocks electrical

           interference

5. List six materials that are good insulators.

   a.   **b.**

   c.   **d.**

   e.   **f.**

6. Draw the schematic symbol for the following switches.
   (*Hint:* Refer to Figure 1–9)

   a. SPST   **b.** SPDT   **c.** DPST   **d.** DPDT

   e. normally open pushbutton   **f.** normally closed pushbutton   **g.** rotary

7. List the number of terminals on each switch.
   (*Hint:* Refer to Figure 1–9.)

Switch	Number of Terminals
a. SPST	
b. SPDT	
c. DPST	
d. DPDT	

8. Draw a simple circuit containing a battery, fuse, lamp, and SPST switch. (*Hint:* Refer to Figure 1–13.)

9. Draw the face of an electrical outlet and indicate the "hot side," ground side, and earth ground connections.

10. List four basic safety procedures that you should observe or practice.

   a.

   b.

   c.

   d.

≡≡≡ SECTION 1–4
EXPERIMENTS

EXPERIMENT 1. Ohmmeter Familiarization

Objective:

To develop essential skills in setting up an ohmmeter for testing electrical/electronic devices.

Introduction:

Usually, an ohmmeter is part of a multimeter and you must be able to select the proper switches in order to place the meter into the resistance or ohmmeter condition. The ohmmeter has its own internal low-voltage source. This voltage is present at the test leads when the meter is in the ohmmeter function. The test leads are placed on the component under test and the amount of current that flows determines the condition of the device as indicated by the meter movement or display.

Materials Needed:

1 Analog or digital multimeter
1 Set of test leads

Procedure:

1. Locate the on/off switch. Turn the meter on.
2. Locate the function switch (this switch determines if the meter is set to measure voltage, current, or resistance). Set this switch to measure resistance or ohms in the R × 1 position.
3. If you are using an analog meter, locate the resistance scale. The pointer should be reading high. If you are using a digital meter, the display will read high or be blinking. The red lead usually goes into the + jack and the black lead into the − jack.
4. Place the test leads into the proper meter jacks, usually marked + and − . Some analog meters may have an ohms adjust control. Allow the meter a few minutes to warm up, and then turn this control until the pointer is at the high end of the resistance scale. Position the pointer directly over the last mark on the scale.
5. Place the tips or ends of the leads together. In other words, short the leads together. The meter should indicate zero. If the analog meter does not show zero, you should be able to turn the ohms adjust (or zero adjust) control until the meter is at zero. Open and close the leads a few times to become familiar with the change in meter indication.
6. Turn off the multimeter.

Fill-in Questions:

1. An ohmmeter measures _____ .

2. Before an ohmmeter is used to test components it must be adjusted to

   _____ .

3. There is a low _____ present at the end of the leads of an ohmmeter.

4. An ohmmeter is usually found in a

   larger unit called a _____ .

EXPERIMENT 2. Testing a Switch

Objective:

To demonstrate how to test a switch in the open and closed condition.

Introduction:

An ohmmeter is connected to the C terminal and the NO or NC terminal of a switch. When the switch is open the resistance between the contacts should be infinite. When the switch is closed the resistance between the contacts should be zero.

Materials Needed:

1 Multimeter
1 Set of test leads
1 Switch

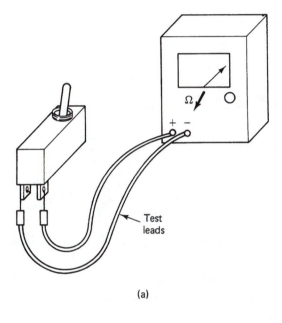

(a)

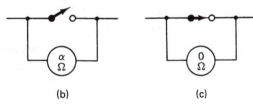

(b)   (c)

Figure 1–20 Testing a switch: (a) pictorial diagram; (b) open switch = infinite ohms; (c) closed switch = zero ohms.

Procedure:

1. Set up the ohmmeter using the procedure described in Experiment 1.
2. Connect the meter leads to the switch as shown in Figure 1–20a.
3. With the switch open the ohmmeter

should indicate infinite resistance (Figure 1–20b). If the resistance is low or zero ohms in either position in which the switch actuator lever is placed, the switch is shorted and should not be used in a circuit.

4. With the switch closed the resistance should read zero ohms (Figure 1–20c). If the resistance reads infinity when the switch actuator lever is placed in either position, the switch is open and should not be used in a circuit.
5. Remove the test leads from the switch.
6. Turn off the multimeter.

Fill-in Questions:

1. An open switch should have _____ ohms of resistance.

2. A closed switch should have _____ ohms of resistance.

3. If a switch being tested reads zero ohms in the open position, the switch is

   _____ .

4. If a switch being tested reads infinite ohms in the closed position, the switch

   is _____ .

EXPERIMENT 3. Testing a Lamp

Objective:

To show the proper method for testing a lamp with an ohmmeter.

Introduction:

Incandescent lamps have fine wire filaments that conduct current and in turn produce light. If these filaments break or burn open, no current can flow and the lamp is considered faulty. An ohmmeter can be used to test the filament of a lamp.

Materials Needed:

1 Multimeter
1 Set of test leads
1 Incandescent lamp

Procedure:

1. Set up the ohmmeter using the procedure described in Experiment 1.

2. Connect or hold the test leads to the lamp as shown in Figure 1–21a. Figure 1–21b shows the electrical symbols for this test.

3. With this connection the ohmmeter should read very low ohms. The filament wires will have some resistance. If the ohmmeter indicates infinite ohms, the lamp is open and cannot be used in a circuit.

4. Turn off the multimeter.

Fill-in Questions

1. An incandescent lamp in good condition will have ____ ohms in its filament wires.

2. If an ohmmeter reads infinite ohms when a lamp is being tested, the lamp is considered _____ .

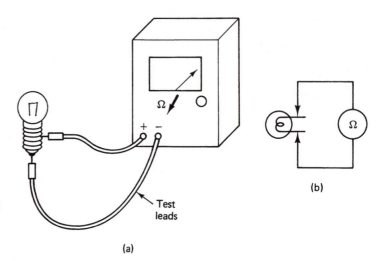

Figure 1–21 Measuring resistance of a lamp: (a) pictorial diagram; (b) schematic diagram.

Test leads

(a)

(b)

≡ SECTION 1–5
INSTANT REVIEW

- *Voltage* is the electrical force that causes current flow in an electrical circuit.
- The unit of measure for voltage is the *volt*.
- *Resistance* is the opposition to current flow in a wire or component.
- The unit of measure for resistance is the *ohm*.
- *Current* is the flow of electrons in an electrical circuit.
- The unit of measure for current is the *ampere*.
- An *electrical circuit* is an electrical path between two or more devices for the purpose of carrying electrical current to produce some useful function.
- An *electrical conductor* has many free electrons and passes current very easily. Its resistance is very low.
- An *electrical insulator* has few free electrons and opposes the flow of current. Its resistance is very high.
- An electrical *plug* attaches to the end of a wire to provide a fast and easy connection.
- An electrical *jack* is the receptacle for an electrical plug. It is usually mounted in a chassis.
- A *schematic symbol* is a line drawing representing the electrical aspects of a device or component.

- An electrical *switch* opens or closes an electrical circuit, or transfers one current-carrying wire from one terminal to another terminal.
- *SPST* stands for "single-pole single-throw switch."
- *SPDT* stands for "single-pole double-throw switch."
- *DPST* stands for "double-pole single-throw switch."
- *DPDT* stands for "double-pole double-throw switch."
- *Pushbutton switches* may have NO and NC contacts. They may be momentary and return to the NC condition when released.
- *Rotary switches* can have several C, NC, and NO contacts mounted on a single wafer. A switch may have several wafers connected together.
- *Ganged together* means that two or more switch contacts move when the actuator lever is activated.
- An *open circuit* does not allow current to flow.
- A *closed circuit* allows current to flow.
- A *short circuit* has a very low resistance and allows a large amount of current to flow. A short circuit is normally avoided.
- An *incandescent lamp* has a thin filament wire that heats up and glows when electrical current flows through it.
- A *neon lamp* glows when a specific voltage is placed on the lamp. A lower voltage will not light the lamp.
- A *LED* (light-emitting diode) is a solid-state device related to diodes. It will glow when the proper voltage and polarity are placed on it.
- An *ohmmeter* measures resistance.
- A *voltmeter* measures voltage.
- An *ammeter* measures current.
- A *multimeter* contains an ohmmeter, voltmeter, and ammeter.
- A *fuse* is a circuit protector that protects against too much current flowing in the circuit.
- A *slow-blow fuse* will allow momentary surge currents without burning open.
- A *circuit breaker* is also a circuit protector, but can be reset with a button or lever.
- An *electrical shock* is a tingling sensation that occurs when electrical current passes through the human body.
- *Electrocution* is death caused by a severe electrical shock.
- A *load* is any device that requires electrical current for operation.
- An *overload* is too many loads, or devices, plugged into a single electrical outlet. Too much current is taken from the power source and could result in a fire.
- *Earth ground* is the third connection of an ac power source to the earth ground.
- A *voltage surge* is a momentary increase in the voltage from an ac power source.
- A *frayed wire* is a wire with the insulated cover broken and the wire exposed.
- *Handle line cords only by the plug end.* Do not pull on the wire to remove a line cord from an electrical outlet.
- *Water is a good conductor* of electrical current.
- *Do not operate* or work with *electrical devices when you are wet or standing in water.*

Circle the most correct answer for each question.

1. Voltage is the:
   a. force that causes current flow
   b. flow of electrons in a wire
   c. opposition to current flow
   d. none of the above

2. The unit for electrical current is the:
   a. volt   b. ohm
   c. ampere   d. load

3. Ohms is associated with:
   a. voltage   b. current
   c. batteries   d. resistance

4. Many free electrons are found in:
   a. an insulator b. a conductor
   c. a switch   d. none of the
                       above.

5. Two wires are made of the same material, have the same diameter, but are of different lengths. The wire with the highest resistance is the:
   a. shortest wire   b. longest wire

6. Two wires are made of same material, have the same length, but their diameters are different. The wire with the highest resistance has the:
   a. largest diameter b. smallest
                           diameter

7. The AWG wire size with the smallest diameter is:
   a. 22   b. 20
   c. 18   d. 10

8. Devices attached to the ends of wire to provide fast and easy connect and disconnect features are called:
   a. jacks   b. fuses
   c. plugs   d. circuit breakers

9. An open circuit is a circuit that has:
   a. current flow   b. no current flow
   c. not enough   d. too much
      voltage   voltage

10. The number of terminals on a DPDT switch is:
    a. 2   b. 4
    c. 6   d. 8

11. A particular circuit has a 4-A fuse. If the current in the circuit increases to 6 A, the fuse will:
    a. short out   b. burn open
    c. remain the same   d. none of the
                            above

12. A multimeter will *not* measure:
    a. voltage   b. resistance
    c. temperature   d. current

13. The amount of electrical current passing through the body that can cause severe shock or even death is:
    a. 0.001 to 0.01 A   b. 0.01 to 0.1 A
    c. 0.1 to 1 A   d. 1 to 10 A

14. If a person was caught by a high-voltage circuit, you could knock the person off the circuit with:
    a. a wooden stick   b. an aluminum
                            pipe
    c. a wet blanket   d. none of the
                            above

15. If a person has been shocked and is unconscious, but you have not been trained in first aid, the first thing you should do is:
    a. use cardiopulmonary resuscitation
    b. cover the victim with blankets or clothing
    c. bandage any burns to the victim's body
    d. call for assistance and then do what you can

16. When too many electrical appliances are plugged into an electrical outlet, the resulting effect could be:

   a. an open circuit   b. a shorted circuit

   c. an overload   d. a., b., and c.

17. One good way to break a line cord is:

   a. to lay it on the floor

   b. to place it next to a refrigerator

   c. to pull on the wire instead of the plug end

   d. none of the above

18. Which of the following is not a conductor of electricity?

   a. plastic   b. water

   c. copper   d. aluminum

19. To operate, most house appliances require only two wires in a line cord. The reason for the third wire is:

   a. to strengthen the line cord

   b. to provide earth ground for safety

   c. to provide a return loop for any built-up charges

   d. none of the above

20. The best type of switch that could be used for safety in the lift lid of a spin-dry washing machine is a:

   a. toggle switch   b. rotary switch

   c. pushbutton switch   d. mercury switch

ANSWERS TO FILL-IN QUESTIONS AND SELF-CHECKING QUIZ

Experiment 1:   (1) resistance   (2) zero   (3) voltage   (4) multimeter

Experiment 2:   (1) infinite   (2) zero   (3) shorted   (4) open

Experiment 3:   (1) low   (2) open

Self-Checking Quiz: (1) a   (2) c   (3) d   (4) b   (5) b   (6) b   (7) a   (8) c
(9) b   (10) c   (11) b   (12) c   (13) b   (14) a   (15) d
(16) c   (17) c   (18) a   (19) b   (20) d

Unit 2

Passive Electronic Devices

INTRODUCTION   *Passive electronic devices* are components used in a circuit that do not amplify or increase the power in the circuit. Passive devices actually restrict the flow of current in a circuit and consume power from the circuit in the form of loads.

UNIT OBJECTIVES   Upon completion of this unit, you will be able to:

1. Identify various types of resistors, capacitors, inductors, transformers, and other miscellaneous components.
2. Explain the resistor color code.
3. Determine the resistance of a resistor by using the color code.
4. Test resistors with an ohmmeter.
5. Explain the operation of a potentiometer.
6. Test a potentiometer with an ohmmeter.
7. Describe the operation of nonlinear resistors.
8. Define the uses of a capacitor.
9. Perform a basic test on a capacitor using an ohmmeter.
10. Define the uses of an inductor.
11. Perform a basic test on an inductor using an ohmmeter.
12. Explain the use of a transformer.
13. Perform a basic test on a transformer using an ohmmeter.
14. Explain the use of miscellaneous passive devices.
15. Define terms associated with passive devices.
16. Use prefixes to units of measure such as *kilo, mega, milli,* and *micro.*

2-1a RESISTORS

All materials have some resistance. In Unit 1 we explained how conductors such as silver and copper have low resistance and offer very little opposition to current; also, that insulators such as glass and rubber have very high resistance and offer a large opposition to current flow. A *resistor* is a specially designed and manufactured electronic component that presents a specifically desired amount of resistance into a circuit.

2-1a.1 The Carbon Resistor

The most commonly used resistor is made of carbon. Carbon is neither a good conductor nor a good insulator, but rather, is in between these two extremes, in the category called *semiconductor*. A composition of carbon granules, an insulating material, and a binder substance are mixed together and shaped into a small solid rod. Wire leads are attached to each end of the rod and then the rod is covered with a nonconductive solid coating. The carbon resistor can have ohmic values of less than 1 Ω to more than 20MΩ.

2-1a.2 Resistor Color Code

Most resistor values are indicated by color bands painted around the resistor, which are referenced to a color code. Figure 2-1 shows the schematic symbol for a resistor and the color code. The colors represent the numbers 0 through 9, powers of 10, tolerance percentages, and failure rates.

The color bands of a resistor are grouped toward one end. To "read" a resistor and calculate its value, the end with the bands is placed to the left. Reading from left to right, the bands indicate:

First color band: first significant number
Second color band: second significant number
Third color band: multiplier
Fourth color band (if any): tolerance
Fifth color band (if any): failure rate

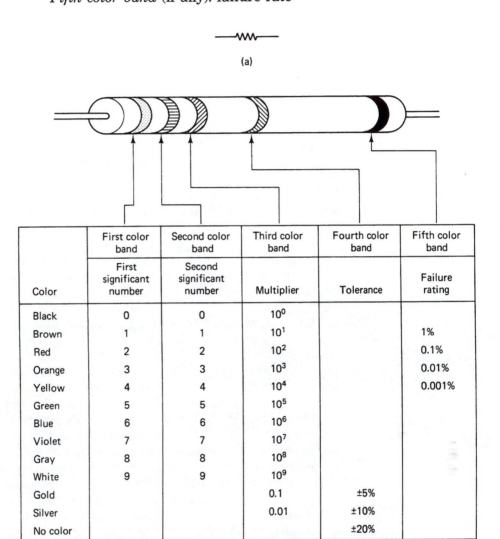

(a)

(b)

Color	First color band	Second color band	Third color band	Fourth color band	Fifth color band
	First significant number	Second significant number	Multiplier	Tolerance	Failure rating
Black	0	0	10^0		
Brown	1	1	10^1		1%
Red	2	2	10^2		0.1%
Orange	3	3	10^3		0.01%
Yellow	4	4	10^4		0.001%
Green	5	5	10^5		
Blue	6	6	10^6		
Violet	7	7	10^7		
Gray	8	8	10^8		
White	9	9	10^9		
Gold			0.1	±5%	
Silver			0.01	±10%	
No color				±20%	

Figure 2–1   Resistor: (a) schematic symbol; (b) color code chart.

For example, if a resistor had the following color bands:

first band = yellow = 4
second band = violet = 7
third band = red = 2 or $\times 10^2$
fourth band = no color = $\pm 20\%$

the resistor would read

$$47 \times 10^2 = 4700 \ \Omega \ \text{at} \pm 20\% = 4.7 \ \text{k}\Omega \ \text{at} \pm 20\%$$

Similarly, if a resistor had the following color bands:

first band = green = 5
second band = blue = 6
third band = gold = 0.1
fourth band = gold = ± 5%

the resistor would read

$$56 \times 0.1 = 5.6 \ \Omega \text{ at } \pm 5\%$$

It is very difficult and expensive to manufacture carbon resistors to an exact value. Most circuits do not require absolute precision; therefore, these types of resistors have a tolerance or range in which they are usable. For example, a 1-kΩ resistor may have a ± 10% tolerance. Ten percent of 1000 is 100; therefore, the actual value of the resistor can vary from 900 to 1100 Ω (1000 − 100 = 900 and 1000 + 100 = 1100).

The failure rating, indicated by the fifth color band on a resistor, is a relatively new specification. It shows the reliability of a resistor in terms of the maximum failure rate in percent per 1000 hours. Carbon resistors are normally required for specialized circuits in military and aerospace applications.

2–1a.3 Resistor Sizes and Power Ratings

In our discussion on conductors, it was shown how resistance was determined by the length, cross-sectional area, and resistivity of the material used. These factors are not necessarily true of resistors, since their resistance is determined by a mixture, not the physical size.

The different physical sizes of resistors relates to their power-handling capability. This is referred to as the wattage rating or the amount of heat that a resistor can dissipate without burning up or altering its internal structure. Normally, if the internal structure of a resistor changes due to overheating, its value will increase. Figure 2–2 shows a comparison of the physical size and power rating of resistors.

Carbon resistors are generally found in sizes of $\frac{1}{8}$ W up to 2 W and are usually color-coded. Resistors with higher power ratings are of different types and normally have their resistance value printed on the side of the resistor. These power ratings can go as high as 250 W.

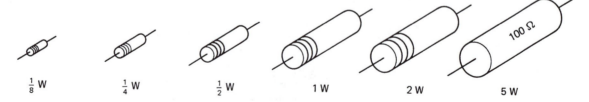

$\frac{1}{8}$ W $\frac{1}{4}$ W $\frac{1}{2}$ W   1 W   2 W   5 W

Figure 2–2 Resistor sizes and power ratings.

2–1a.4 Other Types of Resistors

Other types of fixed resistors are manufactured for specific purposes. The resistance tolerances are usually more precise, on the order of 1% or less; however, they tend to be more expensive.

Wire-wound resistors are copper, nichrome, or other metal alloy wire that is wound around a ceramic core. Leads are attached to each end and the entire unit is sealed in an insulating material. Higher-power-rated resistors are usually wire-wound types. Their main disadvantage is the property of coiled wire and its interaction with the flow of electron current.

Carbon-film resistors appear as regular resistors, but their internal construction is different. A thin resistive film with a mixture of carbon and an insulating material is deposited on a small tubular (rod) ceramic insulating substrate. The desired resistance value is obtained by removing part of the resistive material by cutting a helical pattern along the rod in a spiraling manner. With this technique smaller tolerances ($\pm$ 5% or less) can be achieved and the resistance value is more stable over a wider range of temperatures. The entire unit is then processed with an epoxy coating and the color code is applied.

Deposited-film resistors consist of nichrome, aluminum oxide, tin oxide, or other metal alloys that are heated to high temperatures and a thin film of the material deposited on insulated substrate (support or foundation) material. With some resistors this substrate is cut into a spiraling pattern and placed on a ceramic core. *Thin-film resistors* are classified as having a thickness on the order of 10^{-6} in., whereas *thick-film resistors* are greater than 10^{-6} in.

Some resistors are made into an *integrated circuit* (IC) form and placed into IC packages containing eight or more pins. They are low-power types but take up less space on PC boards.

Precision color-coded resistors are also used in the electronics industry. Their accuracy is $\pm$ 2% or less and can be determined as shown in Figure 2–3a. In this case the third band of color is the third digit of the number, the fourth band is the multiplier, and the fifth band is the tolerance. The colors for the first four bands are the same as with standard color coding, but the tolerance band has some of the original colors assigned different values, as shown.

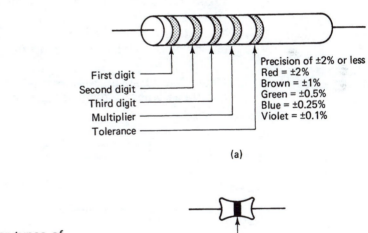

First digit
Second digit
Third digit
Multiplier
Tolerance

Precision of ±2% or less
Red = ±2%
Brown = ±1%
Green = ±0.5%
Blue = ±0.25%
Violet = ±0.1%

(a)

Single black band

(b)

Figure 2–3 Other types of resistors: (a) precision resistor color code; (b) zero-ohm resistor.

The *zero-ohm resistor* is nothing more than a piece of straight wire with a material molded on it which resembles a small resistor. Painted on it is a single black band indicating zero ohms, as shown in Figure 2–3b.

Components are mounted on printed circuit boards using automatic insertion equipment. In some applications a shorting strip made of wire is required between two points to connect two circuits together. In the past, automatic insertion equipment was not able to handle straight wire; therefore, the strips were inserted by hand, which caused delays in the manufacturing process. With the advent of the zero-ohm resistor, automatic insertion equipment was able to handle these shorting strips as normal resistors, which resulted in saving time and money for the manufacturer.

Some resistors are used in circuits that carry large currents; therefore, they must be able to handle more power and are referred to as *power resistors*. These resistors are usually heavier-gage wire-wound types that are covered with a ceramic or porcelain material. They may have a hole through the center to dissipate heat more easily. Various types of fixed resistors are compared in Figure 2–4.

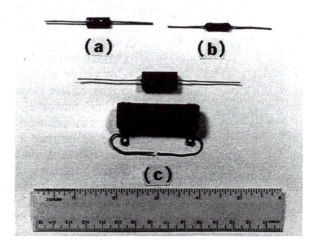

Figure 2–4 Various resistors: (a) carbon color-coded; (b) precision; (c) power.

2–1a.5 Variable Resistors

Variable resistors have some means of mechanically adjusting their resistance. Figure 2–5 shows some common types of variable resistors. A normal resistor has two leads or terminals. A variable resistor has a third terminal, which is movable and indicated by an arrow, as shown, on the schematic symbol of Figure 2–5a.

Figure 2–5 Variable resistors: (a) schematic symbol; (b) wire-wound adjustable type; (c) trimmer type; (d) potentiometer.

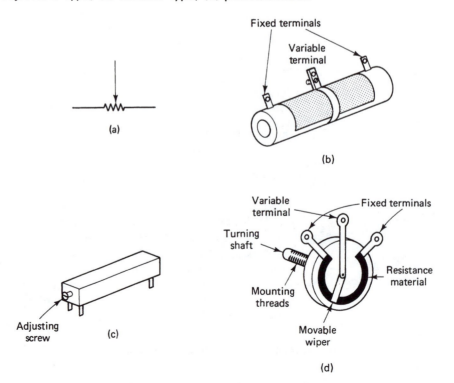

The *wire-wound adjustable resistor* shown in Fibure 2–5b has a section removed from its outer insulating material, exposing the wire. A metal slider can be moved along the wire, which varies the resistance from that variable terminal to the end fixed terminals. This type of resistor is normally of the higher-power type and its resistance is not changed very often. A locking screw is also provided on the variable terminal for securing that terminal to a desired resistance setting.

The *trimmer resistor,* shown in Figure 2–5c, is usually very small and mounts on PC boards. It is normally adjusted by a screwdriver. Trimmer resistors are used to "fine adjust" or set specific resistances to control other circuit functions. They are not adjusted often.

The *potentiometer,* commonly referred to as a *pot,* shown in Figure 2–5d, is the most used variable resistor. The resistance material is usually carbon or copper wire wound on a circular form. A fixed terminal is attached to each end of the resistance material. A variable terminal, also called a *wiper* since it "wipes" across the resistance material, is connected to a shaft that comes out the front of the unit. When the shaft is turned, the resistance changes at the variable terminal in relation to the fixed terminals. The potentiometer is used in a circuit whenever the resistance must be varied frequently. You have used potentiometers countless times. They are used as volume controls, tone controls, balancing controls, and adjusting controls for graphic equalizers in such equipment as stereo music systems, radios, and television receivers.

2–1a.6   Potentiometer Operation

To better understand the operation of a potentiometer, refer to Figure 2–6. If an ohmmeter is placed across the fixed terminals, the total resistance of the pot can be read. If the ohmmeter is placed across the variable terminal and one fixed terminal, the resistance reading will vary when the shaft is rotated. When the wiper moves up, the resistance will increase. When the wiper moves down, the resistance will decrease. If the wiper is placed all the way to the bottom, the ohmmeter will indicate zero ohms. In other words, the leads of the ohmmeter are effectively shorted together. Potentiometers are usually connected where a clockwise motion increases the resistance and a counterclockwise motion decreases the resistance.

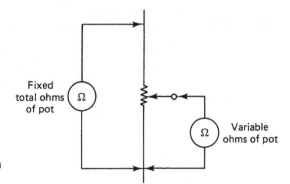

Figure 2–6   Operation of a potentiometer.

The manner in which the resistance is distributed between the fixed terminals is referred to as the *taper* of the pot. A *linear taper* means that the resistance change at the variable terminal is directly proportional to the movement of the wiper. With an *audio* or *logarithmic taper* the resistance changes at a logarithmic rate of base 10 in proportion to the movement of the wiper. This type of pot is used as the volume control in audio circuits because of the nature of the human ear to hear various power levels. With an *S taper,* the resistance increases proportionally with the wiper up to a point, and then as the wiper is increased, the resistance decreases.

A *helical potentiometer*, sometimes called a *helipot*, is a precision potentiometer which requires several turns of the shaft to move the wiper from one fixed terminal to the other. This pot is used in control circuits that require extremely fine adjustment.

There are many variations with pots. Several mounted on the same shaft are referred to as being "ganged together" since the rotation of the shaft varies the resistance of each. Some pots have switches mounted behind them and serve as on/off volume controls for different equipment.

A *rheostat* is a variable resistor with two-terminals: one fixed terminal and one variable terminal. A pot can be wired as a rheostat simply by not connecting one of the fixed terminals. Figure 2–7 shows various types of potentiometers, and Figure 2–8 shows various trimmer resistors.

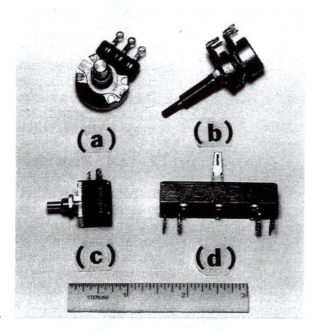

Figure 2–7
Potentiometers: (a) through-chassis with nut mount; (b) double pot; (c) square type; (d) slide type.

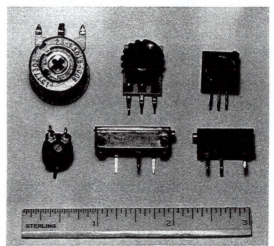

Figure 2–8 Various trimmer resistors.

2–1a.7 Nonlinear Resistors

Nonlinear resistors are devices whose resistance varies on a nonlinear basis when circuit conditions change or with external environmental changes. The majority of these devices have a negative temperature coefficient, although some with a positive temperature coefficient are used in specific circuit applications. Three of these devices are shown in Figure 2-9.

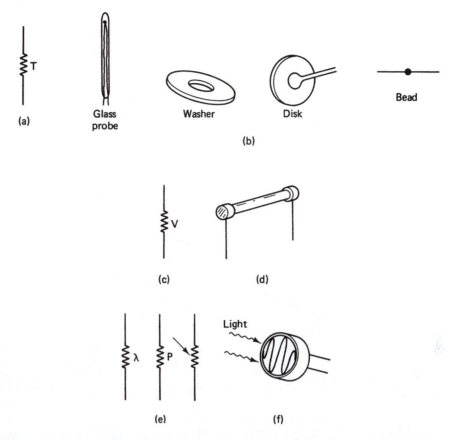

Figure 2-9 Nonlinear resistors: (a) thermistor schematic symbol and (b) physical types; (c) voltage-dependent resistor and (d) physical type; (e) photoresistor schematic symbols and (f) physical type.

The *thermistor* or *thermal resistor* responds to changes in temperature. It can control current flow in a circuit, depending on the temperature of the circuit. Thermistors also find applications in meters where temperature changes are indicated as voltage or current changes.

The *voltage-dependent resistor*, sometimes called a *varistor*, changes resistance with an increase in voltage. In some circuits, this type of resistor is used to oppose voltage surges and protect other components. It may be called a "surgistor" for this application.

The *photoresistor* is made of a semiconductor material: cadmium sulfide (CdS) or cadmium selenide (CdSe). This material is deposited on a ceramic substrate with two attached metal leads and mounted in a case with a window. Its resistance responds to light. When light falls through the window onto the material, electron–hole carriers are released. The current has increased; therefore, it is said that the resistance has decreased. The minimum resistance for a given amount of light is called the *light resistance* (R_{light}) and the maximum resistance, when no light is present, is called the *dark resistance* (R_{dark}). The photoresistor is usually classified as an optoelectronic device. One of its more popular applications is to control the automatic turning on of a lamp when there is no sunlight.

2-1a.8 Units of Measurement

In the electrical/electronics field very large numbers are used to indicate the value of components. These actual numbers are too large to place on components or use on electrical or schematic drawings; therefore, an abbreviated version is used, as shown in Table 2–1. These numbers are based on the powers of 10 as follows:

$$
\begin{aligned}
\text{units} &= 1 = 10^0 \\
\text{tens} &= 10 = 10^1 \\
\text{hundreds} &= 100 = 10^2 \\
\text{thousandths} &= 1000 = 10^3 \\
&\text{and so on}
\end{aligned}
$$

TABLE 2–1

Numerical Values and Units

Numerical Name	Numerical Value	Power of 10	Prefix	Symbol
Trillion	1,000,000,000,000	10^{12}	tera	T
100 billion	100,000,000,000	10^{11}		
10 billion	10,000,000,000	10^{10}		
Billion	1,000,000,000	10^9	giga	G
100 milllion	100,000,000	10^8		
10 million	10,000,000	10^7		
Million	1,000,000	10^6	mega	M
100 thousand	100,000	10^5		
10 thousand	10,000	10^4	myria	my
Thousand	1,000	10^3	kilo	k
Hundred	100	10^2	hecto	h
Ten	10	10^1	deca	da
Unit	1	10^0		
Tenth	0.1	10^{-1}	deci	d
Hundredth	0.01	10^{-2}	centi	c
Thousandth	0.001	10^{-3}	milli	m
10-thousandth	0.0001	10^{-4}		
100-thousandth	0.00001	10^{-5}		
Millionth	0.000001	10^{-6}	micro	μ
10-millionth	0.0000001	10^{-7}		
100-millionth	0.00000001	10^{-8}		
Billionth	0.000000001	10^{-9}	nano	n
10-billionth	0.0000000001	10^{-10}		
100-billionth	0.00000000001	10^{-11}		
Trillionth	0.000000000001	10^{-12}	pico	p
Quadrillionth	0.000000000000001	10^{-15}	femto	f
Quintillionth	0.000000000000000001	10^{-18}	alto	a

Numbers less than 1 are represented as follows:

$$
\begin{aligned}
\text{tenths} &= 0.1 = 10^{-1} \\
\text{hundredths} &= 0.01 = 10^{-2} \\
\text{thousandths} &= 0.001 = 10^{-3} \\
&\text{and so on}
\end{aligned}
$$

The power of the number indicates how far to move the decimal point to express the real number. As an example:

$4.7 \times 10^2 = 4\ \underline{7}\ \underline{0.}$ or $4.7 \times 10^3 = 4\ \underline{7}\ \underline{0}\ \underline{0.}$
decimal moved two places right   decimal moved three places right

In the case of numbers less than 1:

$$4.7 \times 10^{-2} = 0.\underset{\smile}{0} \ \underset{\smile}{4} \ 7 \quad \text{or} \quad 4.7 \times 10^{-3} = 0.\underset{\smile}{0} \ \underset{\smile}{0} \ \underset{\smile}{4} \ 7$$

decimal moved two places   decimal moved three places
left   left

Some powers of 10 used very often are simply written with their symbol; for example:

$$
\begin{aligned}
47{,}000 &= 47 \text{ K} \\
47{,}000{,}000 &= 47 \text{ M} \\
0.047 &= 47 \text{ m} \\
0.000047 &= 47 \ \mu
\end{aligned}
$$

If a resistor was colored yellow, violet, and orange, it could be written as

$$4 \ 7 \text{ plus 3 zeros} = 47{,}000 \ \Omega \quad \text{or} \quad 47 \text{ k} \ \Omega$$

However, to express some numbers in these units you have to count the places to move the decimal. For example, if another resistor were colored yellow, violet, and red, it could be written as

$$4 \ 7 \text{ plus 2 zeros} = 4700 \ \Omega \quad \text{or} \quad 4.7 \text{ k} \ \Omega$$

Notice that to make a whole number smaller using the powers of 10 you have to move the decimal to the left; expressing a number written in powers of 10 to its real value you have to move the decimal to the right.

Numbers less than 1 have a similar relationship. As the powers of 10 becomes more negative, the smaller the real number becomes. For example:

$$
\begin{aligned}
47 \text{ m} &= 47 \times 10^{-3} = 0.047 \\
47 \ \mu &= 47 = 10^{-6} = 0.000047
\end{aligned}
$$

Notice that to express a number less than 1 in powers of 10, the decimal moves to the right, and to express the same number from powers of 10 to the real number, you have to move the decimal to the left. For example:

$$0.068 = \underset{\smile}{0} \ \underset{\smile}{6} \ \underset{\smile}{8} \times 10^{-3} = 68 \text{ m} \quad \text{and} \quad 60 \text{ m} = 68 \times 10^{-3} = 0. \ \underset{\smile}{0} \ \underset{\smile}{6} \ \underset{\smile}{8}$$

Units of measurement commonly used in electronics are shown in Table 2–2

TABLE 2–2
Measurements Commonly Used in Electronics

Power of 10	Prefix	Symbol	Sample Use
10^9	giga	G	gigahertz
10^6	mega	M	megaohm
10^3	kilo	k	kilowatt
10^{-3}	milli	m	milliampere
10^{-6}	micro	μ	microampere
10^{-9}	nano	n	nanosecond
10^{-12}	pico	p	picofarad

2–1b CAPACITORS

Capacitors are electrical devices that can store an electrical charge or voltage temporarily. The capacitor consists essentially of two conducting surfaces separated by an insulating material called a *dielectric,* shown in Figure 2–10b. This dielectric can be air, paper, mica, glass, oil, or plastic film in the form of polystyrene, Mylar, or tantalum.

The *farad* (F) is the unit of capacitance. The size of the plates is one factor that determines the amount of capacitance of a capacitor. A single capacitor with single plates produces very small amounts of capacitance, in the microfarad or picofarad range. To produce larger values of capacitance, the plates are strips of tin foil, with a dielectric between them which also has insulating strips, and are rolled up to form a tubular shape, as shown in Figure 2–10c. The schematic symbol for the capacitor resembles a one-cell battery, as shown in Figure 2–10a. The curved line on one symbol indicates the outer conductor, which is usually connected to the ground potential of a circuit.

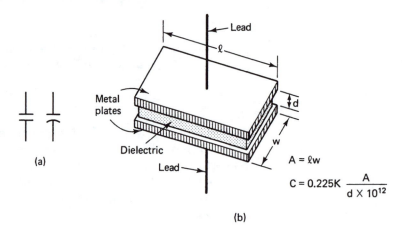

(a)

$$A = \ell w$$

$$C = 0.225K \frac{A}{d \times 10^{12}}$$

(b)

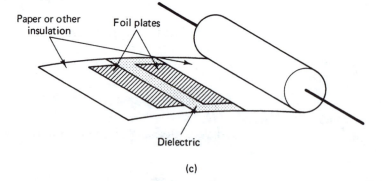

(c)

Figure 2–10 Capacitor: (a) schematic symbols; (b) physical construction; (c) tubular type.

2–1b.1 Types of Capacitors and Their Use

Figure 2–11 shows various types of capacitors. The standard schematic symbol (Figure 2–11a) is associated with those shown in Figure 2–11b–11e. These types of capacitors are used in standard circuits, such as amplifiers, timing, and general circuits. They are used to block dc voltage from one circuit to another, pass an ac signal between circuits, and as timing circuit components. Leads protruding directly out of the ends of a component are referred to as *axial.* Leads coming straight down from a component are called *radial.* Disk capacitors (Figure 2–11e) are thin, flat-looking, disk-shaped devices with radial leads.

Electrolytic capacitors generally use a paste for the dielectric so that

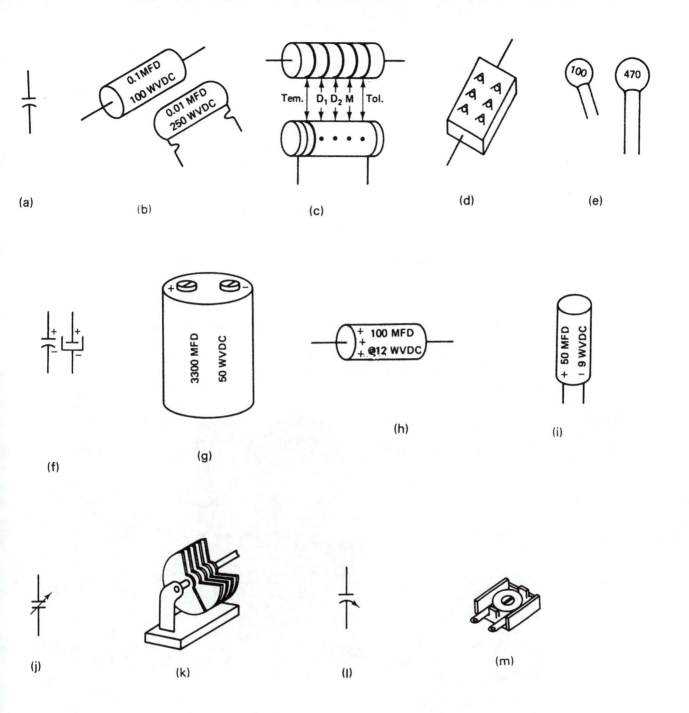

Figure 2–11 Types of capacitors: (a) standard schematic symbol; (b) tubular; (c) color-coded; (d) mica color-coded; (e) disk (f) electrolytic schematic symbol; (g) electrolytic can type; (h) electrolytic axial leads; (i) electrolytic radial leads; (j) variable capacitor schematic symbol; (k) variable (tuning) capacitor; (l) trimmer capacitor schematic symbol; (m) trimmer capacitor.

large amounts of capacitance can be produced in relatively small containers. These types of capacitors are easily identified by the polarity markings on them, as shown by the schematic symbols (Figure 2–11f–i). It is very important that electrolytic capacitors be connected properly in a circuit. The + lead must go to the positive potential and the − terminal must go to the negative potential. Incorrect installation of an electrolytic capacitor can cause it to explode when power is applied to the circuit. For this reason, a *blow hole* with a thin membrane is placed at one end of the capacitor. When the pressure builds up to a certain point the membrane ruptures and the gases escape, which prevents the capacitor from exploding. Large can-type electrolytic capacitors (Figure 2–11g) are generally used as power supply filters for smoothing out unwanted voltage ripple. Other types of electrolytic capacitors (Figure 2–11h and i) can be found in smaller power supplies and general circuits.

Many capacitors have their value printed on them. Also, the amount of voltage they can withstand across their plates is given as a *voltage rating*. For example, a capacitor may read 0.1 μF or mfd (microfarad) at 100 WVDC (working voltage direct current).

A variable capacitor is represented by an arrow across its schematic symbol (Figure 2–11j). One type of air dielectric capacitor that is variable (Figure 2–11k) is used in radio tuning circuits, which help you select one station from another.

A trimmer capacitor (Figure 2–11m) is variable and shows its arrow as one of the plates (Figure 2–11l). Normally, these are very small capacitors with mica as the dielectric which are adjusted for very fine frequency selection.

Figure 2–12 shows various electrolytic capacitors. Figure 2–13 shows other various types of capacitors.

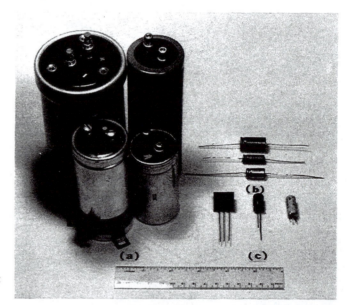

Figure 2–12 Various electrolytic capacitors: (a) large power supply types; (b) tubular with axial leads; (c) radial-lead types.

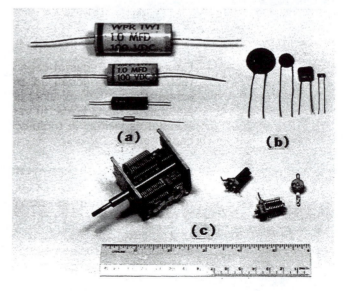

Figure 2–13 Various capacitors: (a) tubular with axial leads; (b) disk type; (c) air dielectric tuning types.

2-1b.2 Capacitor Color Code

There are several methods used to color-code capacitors to indicate their value. Two standard color codes are set by the Joint Army and Navy (JAN) and the Electronic Industries Association (EIA). It would be very difficult to try to remember all the various methods; therefore, it is best to refer to an electronic reference book when using color-coded capacitors. Some of the capacitor color-coding methods are shown in Appendix B.

2-1c INDUCTORS

Inductors are electrical devices that oppose a change in the flow of current and operate on the principle of electromagnetism. Basically, an inductor is a coil of wire, as shown in Figure 2–14e. An air-core inductor

Figure 2–14 Inductors: (a) schematic symbol for air core; (b) schematic symbol for iron core; (c) schematic symbol for adjustable type; (d) schematic symbol for type with taps; (e) coil; (f) power supply type; (g) and (h) radio-frequency types; (i) inductor with taps; (j) adjustable radio-frequency type.

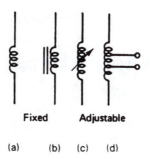

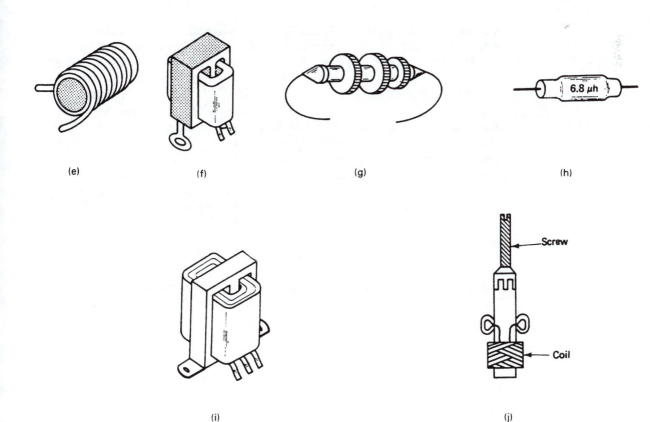

has nothing through its center; the schematic symbol is shown in Figure 2–14a. The *henry* (H) is the unit of inductance. An inductor with an iron core has more inductance than that of an air-core inductor of the same size. The iron-core inductor schematic symbol has two lines beside it, as shown in Figure 2–14b. Inductors used in radio and television circuits are shown in Figure 2–14g and h. Some inductors are variable or adjustable, as indicated by the schematic symbols in Figure 2–14c and d. A tapped inductor is shown in Figure 2–14i and a screw-adjust type is shown in Figure 2–14j. Figure 2–15 shows a comparison of the size of various inductors.

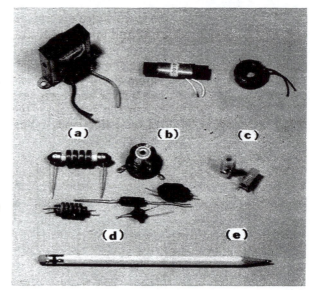

Figure 2–15 Various inductors: (a) power supply type; (b) adjustable iron core; (c) torodial; (d) radio-frequency chokes; (e) adjustable radio-frequency types.

2–1d TRANSFORMERS

A device related to the inductor is the transformer. A *transformer* has at least two coils in close proximity and has the ability to transfer electrical power in one coil to the other coil without being connected together. The transfer of energy is accomplished by electromagnetic induction from one coil to the other, similar to the manner in which two permanent magnets will affect each other. Voltage is applied to the primary winding, marked P, as shown in the schematic symbol of Figure 2–16a. Another voltage is produced at the secondary winding, marked S. A transformer may be a step-up type, where the voltage in the secondary is larger than the voltage in the primary, or a step-down type, where the voltage is less. An iron core, shown in Figure 2–16b, increases the transferring action of a transformer. A transformer may have a center tap (Figure 2–16d and i) or several secondary windings (Figure 2–16e). It may also have several primary windings. A transformer with only three leads is usually an autotransformer (Figure 2–16f), where one lead serves both the primary and secondary windings. Adjustable transformers (Figure 2–16g, l, and m), used in radio and television circuits, change the inductance by moving the iron core in and out of the windings. Figure 2–17 shows a comparison of the size of various transformers.

2–1e MISCELLANEOUS DEVICES

2–1e.1 The Relay

The *relay* is used as an electromechanical switching control device between circuits. There are many types of relays, which can have complex

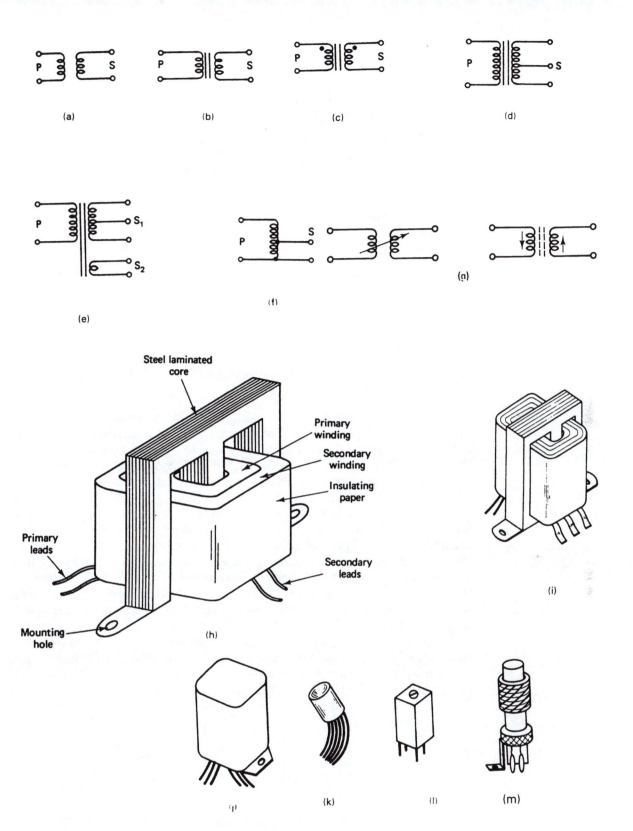

Figure 2-16 Transformers: (a) schematic symbol for air-core type; (b) schematic symbol for iron-core type; (c) schematic symbol showing phasing; (d) schematic symbol for center tap; (e) schematic symbol for multiple windings; (f) schematic symbol for autotransformer; (g) schematic symbol for adjustable types; (h) power transformer; (i) center-tap transformer; (j) in a solid case; (k) miniature type; (l) (m) adjustable types used in radio and television circuits.

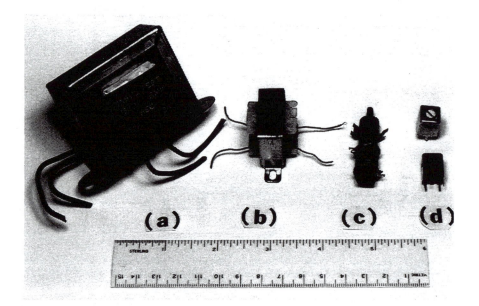

Figure 2–17 Various transformers: (a) power supply types; (b) interstage or output type; (c) tuneable radio-frequency adjustable type; (d) printed-circuit radio-frequency adjustable type.

switching arrangements, but the simplest relay has one coil of wire, an armature, and three electrical contacts, as shown in the schematic symbol of Figure 2–18a. The physical representation of a relay is shown in Figure 2–18b. It has one normally closed (NC) contact and one normally open (NO) contact, with an armature contact that moves between them. The coil of wire has a soft-iron core and is mounted in the relay bracket. Soft iron can easily become an electromagnet when an electric field is applied to it. When the electric field is removed, the soft iron will no longer act as a magnet. The movable armature is mounted above one end of the coil and has a return spring to keep it in the open position normally. The armature contact is touching the normally closed (NC) contact at this time. When voltage is applied to the coil connections, the armature will be attracted toward the end of the coil. The armature contact will be pulled from the NC contact and will now touch the normally open (NO) contact. In other words, the circuit has switched from the NC contact to the NO contact. When the voltage is removed from the coil connections, the armature will be pulled back to its normal position by the return spring. Now the circuit has switched from the NO contact to the

Figure 2–18 Relay: (a) schematic symbol; (b) physical representation.

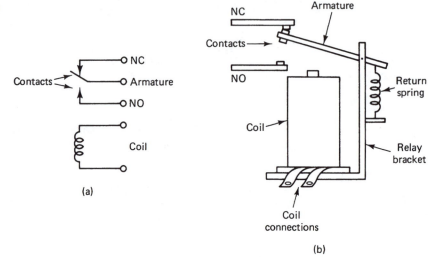

NC contact. Relays are used in automobiles, washing machines, dryers, electronic equipment, and manufacturing equipment. A comparison of the size of various relays is shown in Figure 2–19.

Figure 2–19 Various relays: (a) open type; (b) plastic-enclosed type; (c) mini low-voltage type; (d) reed type.

2–1e.2 Motors

Electrical motors are electromechanical devices used quite extensively in our world. They range in size from very small ones used for hobbies to extremely large ones used in manufacturing. A motor operates on the principle of electromagnetism, and when voltage is applied to it the shaft begins to rotate or turn. A comparison of the size of various motors and the electrical symbol are shown in Figure 2–20.

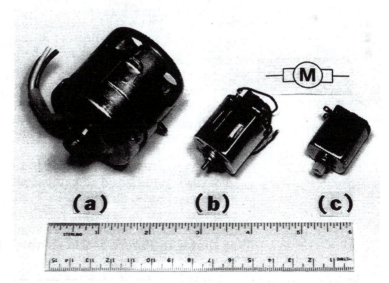

Figure 2–20 Motors: (a) tape drive motor; (b) and (c) hobby motors.

2–1e.3 Audio Devices

Audio pertains to sound. Certain audio devices are used in radios, television sets, tape recorders, and other sound equipment. Figure 2–21 shows three commonly used audio devices.

The *microphone* (Figure 2–21a) is used to transfer sound waves into electrical impulses that are input to an amplifier or other electronic

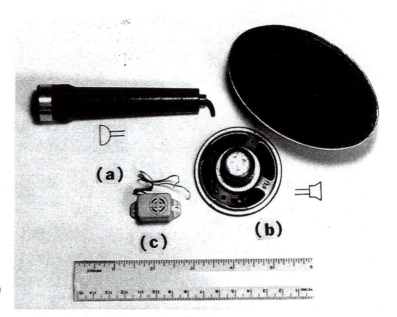

Figure 2–21 Audio devices: (a) microphone; (b) loudspeakers; (c) buzzer.

device. The microphone is referred to as an *input transducer* because it changes one form of energy into another form of energy to be used in an electronic system.

The *loudspeaker* (Figure 2–21b), sometimes referred to simply as a speaker, is an *output transducer* that transfers electrical impulses into sound waves. Speakers come in various sizes, from over 1 ft in diameter to a few tenths of an inch in diameter.

A new audio component used is computers and other small circuits as an alarm or indicator is the *buzzer* (Figure 2–21c). A buzzer may be located a considerable distance from a circuit or it may be mounted on the circuit board.

2–1e.4 Circuit Indicators

The specific action of electronic circuits is not visible to the human eye; therefore, special devices must be placed on the circuit at special points to serve as circuit indicators. A *meter* is a device for measuring or indicating circuit conditions. You have already learned to use an ohmmeter to test switches and lamps. Other meters, such as a voltmeter or ammeter (measures current), may be connected directly into a circuit on the circuit board. Figure 2–22a shows the schematic symbol and a real meter.

The *seven-segment display* is a numerical output indicator often used in digital meters and in computers and other circuits. The display has seven segments, as shown in Figure 2–22b, which light up in various combinations to produce numbers and some letters. Seven-segment displays can be manufactured together to produce a multidigit display, as shown in Figure 2–22c. With this arrangement more numbers can be seen at one time.

2–1e.5 Other Devices

The *piezoelectric crystal* is a specially cut mineral that vibrates when a voltage is placed on it. Crystals are used in radio, television, computer, medical, and various other circuits to provide very accurate frequencies. The actual crystal is mounted in a special case, as shown in Figure 2–23a.

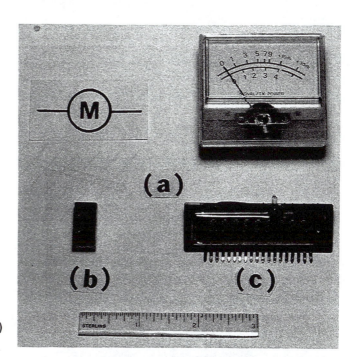

Figure 2–22 Circuit indicators: (a) meter; (b) seven-segment display; (c) multidigit display.

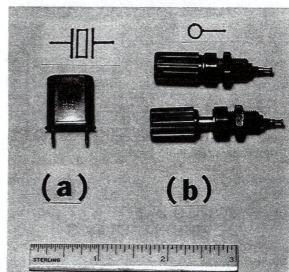

Figure 2–23 Other devices: (a) crystal; (b) binding posts.

Many electronic devices must have features that permit a temporary connection of wires or test leads to other electronic devices, such as meters and test equipment. A *binding post,* shown in Figure 2–23b, is used to make temporary wire connections and is mounted in the chassis or cabinet of electronic equipment. Wires with banana-type plugs can be pushed into the end of the binding post to make a connection. Regular wires with the insulation removed on the ends can be used. The knurl piece of the binding post is unscrewed to reveal a hole in the shaft. The wire is placed into the hole and the knurl piece is tightened back on the wire.

An *antenna* is an exposed conductor that radiates or intercepts electromagnetic waves. The antenna is a very important component for transmitting and receiving radio and television signals. The antenna is used to convert electrical signals into electromagnetic waves for broadcast into the atmosphere at the transmitter and is also used to convert the electromagnetic waves back into electrical signals at the receivers. Figure 2–24 shows its schematic symbol and some familiar uses.

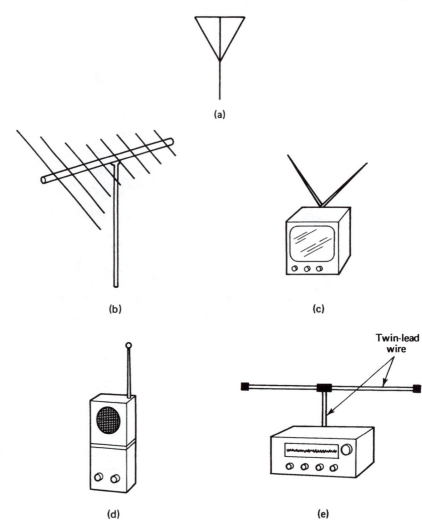

Figure 2–24 Antenna: (a) schematic symbol; (b) outside TV type; (c) indoor rabbit ears; (d) whip type; (e) twin-lead FM type.

≡≡≡≡SECTION 2–2
DEFINITION EXERCISES

Write a brief description of each of the following terms.

1. Passive device _____

_____

2. Carbon resistor _____

_____

3. Wire-wound resistor _____

_____

4. Deposited-film resistor _____

_____

5. Power rating _____

_____

6. Wattage rating _____

_____

7. Zero-ohm resistor _____

_____

8. Color code _____

_____

9. Wire-wound adjustable resistor _____

_____

10. Potentiometer _____

_____

11. Trimmer resistor _____

_____

12. Helical potentiometer _____

_____

13. Rheostat _____

_____

14. Thermistor _____

_____

15. Voltage-dependent resistor _____

_____

16. Photoresistor _____

_____

17. Powers of 10 _____

_____

18. Capacitor _____

_____

19. Farad _____

_____

20. Axial leads _____

_____

21. Radial leads _____

_____

22. Electrolytic capacitor _____

_____

23. Blow hole _____

_____

24. Variable capacitor _____

_____

25. Trimmer capacitor _____

_____

26. Inductor _____

_____

27. Choke _____

_____

28. Henry _____

_____

29. Transformer _____

_____

30. Primary winding _____

_____

31. Secondary winding _____

_____

32. Step-up transformer _____

_____

33. Step-down transformer _____

_____

34. Relay _____

_____

35. Armature _____

_____

36. Normally open contacts _____

_____

37. Normally closed contacts _____

_____

38. Audio _____

_____

39. Microphone _____

_____

40. Transducer _____

_____

41. Loudspeaker _____

_____

42. Buzzer _____

_____

43. Meter _____

_____

44. Seven-segment display _____

_____

45. Multidigit display _____

_____

46. Piezoelectric crystal _____

_____

47. Binding post _____

_____

48. Antenna _____

_____

49. Mega _____

_____

50. Kilo _____

_____

51. Milli _____

_____

52. Micro _____

_____

53. Pico _____

_____

≣ SECTION 2-3
EXERCISES AND PROBLEMS

Complete this section before beginning the next section.

1. Draw the schematic symbol for a fixed resistor.

2. Draw the physical shape of a carbon resistor.

3. Draw the schematic symbol of a potentiometer.

4. Sketch the physical shape of a potentiometer.

5. Write the appropriate number of the resistor color code after each color.

		Tolerances
a. black ____	**g.** blue ____	
b. brown ____	**h.** violet ____	**m.** gold ____
c. red ____	**i.** gray ____	**n.** silver ____
d. orange ____	**j.** white ____	**o.** none ____
e. yellow ____	**k.** gold ____	
f. green ____	**l.** silver ____	

6. Express the following real numbers as a number between 1 and 10 using the powers of 10. *Examples:* $47,000 = 4.7 \times 10^4$, $0.0025 = 2.5 \times 10^{-3}$.

a. 16,900 = _____   b. 3800 = _____

c. 220 = _____   d. 45,000,000 = _____

e. 69.4 = _____   f. 364.2 = _____

g. 0.006 = _____   h. 0.0000038 = _____

i. 0.0726 = _____   j. 0.00000052 = _____

7. Express the following powers-of-10 numbers as their real values.
   Examples: $2.5 \times 10^2 = 250$, $5.4 \times 10^{-4} = 0.00054$.

   a. $119 \times 10^2 =$ _____   b. $3.3 \times 10^3 =$ _____

   c. $74.2 \times 104 =$ _____   d. $0.58 \times 10^3 =$ _____

   e. $0.067 \times 10^2 =$ _____   f. $475 \times 10^{-3} =$ _____

   g. $24 \times 10^{-4} =$ _____   h. $78.5 \times 10^{-3} =$ _____

   i. $0.017 \times 10^{-2} =$ _____   j. $0.009 \times 10^{-3} =$ _____

8. Express the following numbers to 10^3 using the symbol k.
   Examples: $4700 = 4.7$ k, $47,000 = 47$ k, $470,000 = 470$ k.

   a. 1000 = _____   b. 1800 = _____

   c. 2200 = _____   d. 3300 = _____

   e. 5600 = _____   f. 6800 = _____

   g. 10,000 = _____   h. 22,000 = _____

   i. 33,000 = _____   j. 56,000 = _____

   k. 68,000 = _____   l. 100,000 = _____

   m. 220,000 = _____   n. 330,000 = _____

   0. 560,000 = _____   p. 680,000 = _____

9. Express the following numbers to 10^6 using the symbol M.
   Examples: $470,000 = 0.470$ M, $47,000,000 = 47$ M.

   a. 1,000,000 = _____   b. 1,200,000 = _____

   c. 2,200,000 = _____   d. 3,300,000 = _____

   e. 4,700,000 = _____   f. 5,600,000 = _____

   g. 7,800,000 = _____   h. 10,000,000 = _____

   i. 15,000,000 = _____   j. 100,000,000 = _____

   k. 330,000 = _____   l. 220,000 = _____

10. Express the following numbers to 10^{-3} using the symbol m.
Examples: 0.004 = 4 m, 0.0045 = 4.5 m.

   a. 0.001 = _____   b. 0.0015 = _____

   c. 0.005 = _____   d. 0.015 = _____

   e. 0.0006 = _____   f. 0.000032 = _____

11. Express the following numbers to 10^{-6} using the symbol μ.
Examples: 0.000004 = 4 μ, 0.0000045 = 4.5 μ.

   a. 0.000001 = _____   b. 0.0000066 = _____

   c. 0.0000007 = _____   d. 0.000080 = _____

   e. 0.00000002 = _____   f. 0.000000003 = _____

   g. 0.0000000056 = _____ h. 0.004200 = _____

12. For the following numbers, list their power of 10, prefix, and symbol. (Refer to Table 2–1.)

Number	Power of 10	Prefix	Symbol
1,000,000,000			
1,000,000			
1000			
0.001			
0.000001			
0.000000001			
0.000000000001			

13. Calculate the value of the following resistors from the band of colors given. The first band is on the far left. *Example:* yellow violet red silver   4700 Ω at 10%

   a. red red orange silver   _____

   b. brown black red gold   _____

   c. yellow violet black   _____

   d. orange orange red gold   _____

   e. blue gray orange silver   _____

   f. brown black black gold   _____

   g. green black green   _____

   h. orange orange gold silver   _____

   i. green blue brown gold   _____

   j. brown gray red   _____

k. orange orange brown gold   _____

l. red red green silver   _____

m. blue gray yellow silver   _____

n. yellow violet orange gold   _____

o. white brown brown gold   _____

p. gray red orange gold   _____

q. brown green green silver   _____

r. yellow yellow orange gold   _____

s. orange white yellow   _____

t. brown gray brown gold   _____

u. green blue red   _____

v. brown red red silver   _____

w. blue gray silver gold   _____

14. List the color of each band for the resistors given.

	Resistor	First	Second	Third	Fourth
			Band Color		
a.	2.2 kΩ at 20%	_____	_____	_____	_____
b.	4.7 Ω at 5%	_____	_____	_____	_____
c.	560 Ω at 10%	_____	_____	_____	_____
d.	180 kΩ at 20%	_____	_____	_____	_____
e.	12 kΩ at 5%	_____	_____	_____	_____
f.	0.33 Ω at 5%	_____	_____	_____	_____
g.	2.7 Ω at 5%	_____	_____	_____	_____
h.	10 kΩ at 10%	_____	_____	_____	_____
i.	470 kΩ at 20%	_____	_____	_____	_____
j.	680 Ω at 20%	_____	_____	_____	_____
k.	15 kΩ at 10%	_____	_____	_____	_____
l.	750 kΩ at 5%	_____	_____	_____	_____
m.	5.6 MΩ at 10%	_____	_____	_____	_____

	Band Color			
Resistor	First	Second	Third	Fourth
n. 39 kΩ at 20%	_____	_____	_____	_____
0. 10 MΩ at 5%	_____	_____	_____	_____

15. Calculate the resistance value and tolerance for each resistor. Express the resistance in kilohms or megohms where applicable: for example, 2.2 kΩ, 22 kΩ, 220 kΩ, 3.3 MΩ, and so on.

	Band Color					
Resistor	First	Second	Third	Fourth	Ω	at %
a	Red	Red	Brown	Gold	_____	_____
b	Yellow	Violet	Orange	Silver	_____	_____
c	Blue	Gray	Yellow	None	_____	_____
d	Orange	Orange	Red	Silver	_____	_____
e	Green	Blue	Gold	Gold	_____	_____
f	Brown	Black	Blue	None	_____	_____
g	Red	Violet	Brown	Silver	_____	_____
h	Brown	Black	Yellow	Gold	_____	_____
i	Orange	White	Brown	Gold	_____	_____
j	Brown	Black	Orange	Silver	_____	_____
k	Brown	Black	Gold	Gold	_____	_____
l	Yellow	Violet	Silver	Gold	_____	_____
m	Brown	Green	Red	Silver	_____	_____
n	Gray	Red	Gold	Gold	_____	_____
o	Brown	Black	Red	None	_____	_____

16. Draw the schematic symbol for the following nonlinear resistors.

a. thermistor   **b.** voltage-dependent resistor   **c.** photoresistor

17. Draw the schematic symbol and physical appearance for the following types of capacitors.

    a. tubular   **b.** disk   **c.** electrolytic

    d. tuning   **e.** trimmer

18. Draw the schematic symbol for the following components.

    a. air-core inductor   **b.** iron-core inductor

    c. variable inductor   **d.** air-core transformer

    e. iron-core transformer   **f.** center-tapped transformer

    g. variable transformer   **h.** relay

    i. crystal   **j.** motor

    k. microphone   **l.** loudspeaker

    m. buzzer   **n.** meter

    o. binding post   **p.** antenna

19. Match the term in column A with its proper description in column B.

Column A

_____ **a.** Resistor

_____ **b.** Potentiometer

_____ **c.** Thermistor

_____ **d.** Photoresistor

_____ **e.** Voltage-dependent resistor

_____ **f.** Ohm

_____ **g.** Capacitor

_____ **h.** Farad

_____ **i.** Inductor

_____ **j.** Henry

_____ **k.** Transformer

_____ **l.** Color code

Column B

1. unit of inductance
2. opposes current flow
3. changes one voltage to another voltage
4. unit of resistance
5. changes with temperature
6. used to find values
7. stores a charge temporarily
8. changes value with voltage change
9. unit of capacitance
10. adjustable resistance
11. changes with light
12. a coil of wire

20. Match the component in column A with its proper description in column B.

Column A

_____ **a.** Relay

_____ **b.** Motor

_____ **c.** Microphone

_____ **d.** Speaker

_____ **e.** Buzzer

_____ **f.** Seven-segment display

_____ **g.** Meter

_____ **h.** Multidigit display

_____ **i.** Crystal

_____ **j.** Binding post

Column B

1. produces audio sounds
2. vibrates when voltage is placed on it
3. indicates circuit voltage
4. converts sound to electrical pulses
5. isolates one circuit from another
6. used in alarm circuits
7. temporary connection for wires
8. rotates when voltage is applied
9. numerical indicator
10. shows several numbers at one time

EXPERIMENT 1. Testing Resistors

Objective:

To develop skills in correctly determining resistor values using the color code and testing resistors with an ohmmeter.

Introduction:

When measuring resistance, you must be able to read the ohms scale correctly, as shown in Figure 2-25. The ohms scale is the top one shown on the analog meter face. Pay attention to this scale and ignore the others at this time. Notice that this scale reads from right to left, zero on the right and ascending numbers as it progresses left.

The function/range switch is set on 1 k (k stands for "kilo" and means 1000). This means that the numbers on the ohms scale represent thousands: for example, 2 = 2000, 5 = 5000, 10 = 10,000, and so on. In other words, the range switch acts as a multiplier of the numbers read on the scale. Other ranges include

$$1 = \times 1$$
$$100 = \times 100$$
$$10 \text{ k} = \times 10,000$$
$$1 \text{ M} = \times 1,000,000$$

(M, mega, stands for 1 million)

The tip of the pointer is resting between 50 and 70 of the ohms scale. Midway between 50 and 70 is a large line indicating 60; however, the pointer is resting on a smaller line halfway between 50 and the 60 line, which would indicate 55. Therefore, 55 times the range switch setting of 1 k equals 55 kΩ or 55,000 Ω. If the range switch were set on 10 kΩ, 55 times 10 k would indicate a meter reading of 550 kΩ or 550,000 Ω. Similarly, if the range switch were set on 1, 55 the quantity times 1 would indicate a meter reading of 55 Ω.

Notice the right side of the ohms scale. Halfway between 0 and 2 there is a large line indicating 1. There are four smaller division lines indicating values less than 1. Starting from 0, it takes five total lines to reach line 1. Therefore, 1 divided by 5 will indicate that each small line is equal to 0.2. If the pointer was resting on the third line from 0, the reading would be 3 times 0.2 equals 0.6. With the range switch set on 100, the meter would be indicating that 0.6 times 100 equals 60 Ω.

A simple formula can be used to remember how to find the value of a subdivision on any meter scale:

value of each subdivision =
$$\frac{\text{larger number} - \text{smaller number}}{\text{total subdivisions between numbers}}$$

As an example, from number 5 to number 10, it takes 10 subdivisions. The difference be-

Figure 2-25 Reading ohms (range switch set at 1 k pointer indicates 55 kΩ).

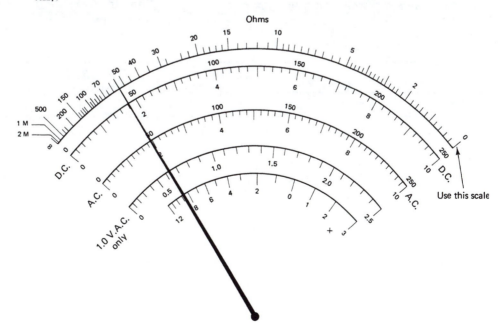

tween 5 and 10 is 5; therefore, 5 divided by 10 indicates that each subdivision is equal to 0.5

value of each subdivision =

$$\frac{10-5}{10} = \frac{5}{10} = 0.5$$

If the pointer was resting 3 subdivisions from the number 5 toward the number 10, the meter reading would be 3 times 0.5 equals 1.5 plus the number 5, for a total of 6.5. With the range switch set on 100, the meter would be indicating that 6.5 times 100 equals 650 Ω. Follow this procedure any time that the pointer falls between any given numbers on the ohms scale.

The most accurate ohm readings occur near the middle of the scale. When making a resistance measurement, always try to get the pointer toward this area by changing the range switch.

Materials Needed:

Six carbon-composition resistors, suggested values:

1	47 Ω	1	22 kΩ
1	220 Ω	1	470 kΩ
1	1 kΩ	1	1 MΩ

Equipment Required:

1 Multimeter or DVM

Procedure:

1. Take each resistor and read its color code.

2. List the colors in the following table as indicated. You may want to use the following abbreviations for the colors:

Black	= Blk		Brown	= Brn
Red	= Red		Orange	= Org
Yellow	= Yel		Green	= Grn
Blue	= Blu		Violet	= Vio
Gray	= Gra		White	= Wht
Silver	= Sil		Gold	= Gol
None	= No			

3. Calculate the resistance value and tolerance and enter them in the table at the appropriate places.

4. Calculate the tolerance of the resistor in terms of ± ohms and list in the table. As an example:

4.7 kΩ at ± 10% = 4700 × 0.1 = ±470 Ω

5. Set the ohmmeter on the proper range. Zero-adjust the meter for that range. *Each time you change a range, be sure to zero-adjust the meter.*

6. Figure 2–26a shows the schematic representation for placing the meter across a resistor to be measured. Place the meter across the resistor as shown in Figure 2–26b. Read the meter and enter this value into the table at the appropriate place.

7. Compare the measured resistance value with the calculated resistance value and the ± tolerance. Indicate by "yes" or "no" in the table if the resistor is within tolerance.

Resistor	Color band				Calculated Ohms (%)	Calculated Tolerance (± Ω)	Measured Ohms	Within Tolerance?
	First	Second	Third	Fourth				
a	___	___	___	___	___	___	___	___
b	___	___	___	___	___	___	___	___
c	___	___	___	___	___	___	___	___
d	___	___		___	___	___	___	___
e	___	___	___	___	___	___	___	___
f	___	___	___	___	___	___	___	___

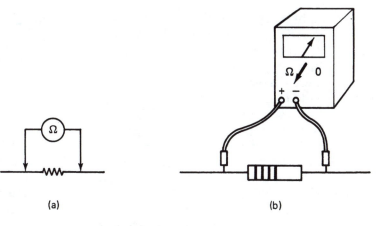

Figure 2–26 Measuring the ohmic value of a resistor: (a) schematic representation; (b) actual ohmmeter across resistor.

(a)

(b)

For example:

Measured resistance = 4850 Ω
Calculated resistance = 4700 Ω
± tolerance = + 470 Ω
5170 Ω

Is 4850 Ω within or below 5170 Ω? Yes.

Note! If the measured resistance value is greater than the calculated resistance value, add the tolerance value; however, if it is less, subtract it.

8. Measure all of the resistors as indicated in steps 5, 6, and 7.

Fill-in Questions:

1. When using an ohmmeter, the first thing to do is to _____ adjust the meter.

2. When measuring the resistance of a resistor, be sure to set the ohmmeter to the proper _____ .

3. A resistor to be used in a circuit should be within the _____ indicated on the resistor.

4. The physical size of resistors is determined by their _____ rating.

EXPERIMENT 2. Testing Capacitors

Objective:

To develop skills for a simple test on capacitors using an ohmmeter.

Introduction:

There are four possible causes for a malfunctioning capacitor. It can become open, shorted, leaky, or have a change of value in the capacitance. The best way to check capacitors is with the use of a capacitor checker, an instrument that is relatively easy to use. You select the capacitance range of the capacitor, connect the capacitor to the leads, and the display will indicate the exact capacitance. This instrument may also test the capacitor for current leakage between the plates. In many cases a capacitor checker is not available; therefore, a multimeter or DVM can be used to perform a go/no go test on a capacitor. Figure 2–27 shows how to test a capacitor with a multimeter.

Using an ohmmeter, you can test a capacitor for an open, short, and leaky condition. As the ohmmeter is initially

Figure 2–27 Testing a capacitor: (a) ohmmeter connection; (b) reverse ohmmeter connection; (c) voltmeter connection.

Testing a

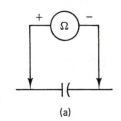

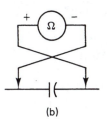

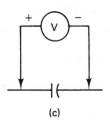

(a)

(b)

(c)

connected to a capacitor, the pointer should deflect or the digital display will indicate a momentary current flow. When the capacitor charges up to the voltage across the ohmmeter leads, current stops flowing and the meter should indicate infinity (Figure 2–27a). If the meter leads are reversed and placed on the capacitor (Figure 2–27b), the pointer deflection will be nearly twice as much as before, because the voltage charge on the capacitor is now in series with the voltage of the meter leads. With a DVM, there is no visible deflection of a pointer, but a capacitor can be checked to see if it will store a charge.

An alternate test would be to place the ohmmeter across the capacitor leads (Figure 2–27a). After the capacitor is charged, remove the ohmmeter. Set the meter to read low dc volts and place it back on the capacitor leads with the same polarity (Figure 2–27c). The meter should indicate the voltage of the charged capacitor, and then it will decrease as the capacitor discharges through the meter. The malfunction of a capacitor can be determined thus:

> *Open:* no meter deflection or reading (the capacitor will not store a charge).
> *Short:* the meter will indicate zero resistance.
> *Leaky:* the meter will indicate some resistance less than infinity.

Large-value electrolytic capacitors will take a long time to charge up to the total voltage on the ohmmeter leads, so you will have to be patient in testing these capacitors. In some cases new electrolytic capacitors have excessive leakage current. This leaky condition may continue for some time after the capacitor is placed into a working circuit. Eventually, this leakage current becomes minimum and the circuit will operate properly. In other words, an electrolytic capacitor has to be "conditioned" or "ripened" in the circuit.

Materials Needed:

Four capacitors, with suggested values:

1 0.1 μF   *1* 10.0 μF
1 1.0 μF   *1* 100.0 μF

Equipment Required:

1 Multimeter or DVM

Procedure:

1. Adjust the ohmmeter and set it to a middle range.
2. Place the meter leads across the 100-μF capacitor as shown in Figure 2–27a. The meter should go to zero and then begin to count up or increase. When it reaches infinity, remove the meter leads. The capacitor is now storing a charge.
3. Set the meter on the dc volts function. Connect the negative or black lead of the meter to the negative lead of the capacitor. Connect the positive or red lead of the meter to the positive lead of the capacitor, as shown in Figure 2–27c. The meter should increase to some value and then rapidly decrease to zero as the capacitor discharges through the meter.
4. Test all of the capacitors using steps 2 and 3.

Notice that the smaller-value capacitors are more difficult to use in reading an indication, because they charge and discharge at a very fast rate.

Fill-in Questions:

1. When an ohmmeter is placed across a capacitor, it will _____ up to the voltage at the test leads.

2. When a voltmeter is placed across a charged capacitor, it will _____ through the meter.

3. The charging effects of small-value _____ are difficult to see on an ohmmeter.

4. If an ohmmeter reads zero continually when placed across a capacitor, it may be _____ .

5. If an ohmmeter reads infinity continually when placed across a capacitor, it may be _____ .

EXPERIMENT 3. Testing Inductors and Transformers

Objective:

To demonstrate how to test inductors and transformers for open and shorted windings.

Introduction:

The coil of an inductor or the windings of a transformer can be tested for an open by making a continuity test with an ohmmeter. If the dc resistance of the winding is known, you can make a resistance check for a partial short. In either case, if a problem is found, the component will not perform properly in a circuit.

Materials Needed:

Several inductors and transformers.

Equipment Required:

Standard ohmmeter or DVM.

Procedure:

1. Adjust the ohmmeter and set it to a low range, perhaps R × 1.
2. Place the ohmmeter leads across the leads of an inductor, as shown in Figure 2–28a. Record the meter reading here ____ . The meter should indicate low ohms. If the meter shows zero, there could be a short in the windings of the coil. If the meter indicates infinity, the coil is open.
3. Remove the ohmmeter leads from the inductor.
4. Place the ohmmeter leads across the leads of the primary winding of a transformer as shown in Figure 2–28b. Record the meter reading here ____ . If the resistance of the windings is known, you can determine whether the winding is okay or defective.
5. Place the ohmmeter leads across the leads of the secondary winding of the transformer. Record the meter reading here ____ . If the secondary winding has a center tap, you must check this lead with the other leads. The resistance from the center tap to either end of the secondary winding should be about one-half of the resistance from the end-to-end lead meter reading. Be sure to test all windings.

Fill-in Questions:

1. The meter reading of a good winding of an inductor or transformer should read

   ____ ohms.

2. If the meter reading of a winding is zero

   ohms, the winding is probably _____

   _____ .

3. If the meter reading of a winding is infinite ohms, the winding is _____

   _____ .

Figure 2–28 Testing windings: (a) inductor; (b) transformer.

(a)

(b)

⎯⎯⎯SECTION 2–5 INSTANT REVIEW

- An electronic *passive device* does not amplify or increase the power in a circuit.
- The *carbon resistor* is the most commonly found resistor.
- *Wire-wound resistors* are usually the precision type.

- *Deposited-film resistors* are usually the precision type.
- The *power rating* of a resistor indicates how much heat it can dissipate; the more power, the larger resistor needed.
- The unit of power is the *watt*.
- The *wattage rating* is the same as the *power rating*.
- A *zero-ohm resistor* provides a link or connecting strip.
- The *color code* indicates the value of electronic components.
- *Wire-wound adjustable resistors* are used in semi-fixed circuits. They are usually not moved often.
- A *potentiometer* is an adjustable resistor used to control voltage or current, such as a radio volume control.
- A *trimmer resistor* is a small adjustable resistor used to fine adjust a circuit where a precise resistance is needed.
- A *helical potentiometer* is an adjustable resistor with many revolutions to produce very accurate resistances.
- A *rheostat* is similar to a potentiometer but has only two terminals and is used to control current.
- A *thermistor* is a temperature-sensitive resistor. With an increase in temperature the resistance decreases, and vice versa.
- A *voltage-dependent resistor* changes resistance with changes in voltage.
- A *photoresistor* is sensitive to light. With an increase in light the resistance decreases, and vice versa.
- *Powers of 10* provide a short method of expressing large numbers based on the exponent or power of the numbers.
- *Mega* (M) stands for 1,000,000 or 10^6.
- *Kilo* (k) stands for 1000 or 10^3.
- *Milli* (m) stands for 0.001 or 10^{-3}.
- *Micro* (μ) stands for 0.000001 or 10^{-6}.
- *Pico* (p) stands for 0.000000000001 or 10^{-12}.
- A *capacitor* is an electronic component capable of storing a temporary voltage charge.
- The *farad* is the unit of capacitance.
- *Axial leads* come directly out of each end of a component and point in opposite directions.
- *Radial leads* come out the same side or end of a component and point in the same direction.
- An *electrolytic capacitor* is a special capacitor marked with polarity signs which indicate how it is to be connected in a circuit.
- The *blow hole* on an electrolytic capacitor will release gas pressures if it is connected incorrectly in a circuit.
- A *variable capacitor* is adjustable, has a dielectric of air or mica, and is used in tuning radio circuits.
- A *trimmer capacitor* is an adjustable capacitor for fine adjustment control for precise frequency adjustment.
- An *inductor* is a coil of wire that opposes a change in current in a circuit.
- A *choke* is another name for an inductor generally used in radio circuits.
- *Henry* is the unit of inductance.
- A *transformer* can transfer the power in its primary winding to its secondary winding without a physical connection.

- Voltage is applied to the *primary winding* of a transformer.
- A larger or smaller voltage is induced in the *secondary winding* of a transformer.
- A *step-up transformer* increases the voltage at its secondary winding.
- A *step-down transformer* decreases the voltage at its secondary winding.
- A *relay* is an electromechanical control device that isolates two different circuits.
- The *armature* is the movable part of a relay.
- The *normally open contacts* are open when the relay is not energized, but close when the armature is energized.
- The *normally closed contacts* are closed when the relay is not energized, but open when the armature is energized.
- *Audio* pertains to sound.
- A *microphone* converts sound energy to electrical energy.
- A *transducer* is any device that converts one form of energy into another form of energy.
- A *loudspeaker* converts electrical energy into sound energy.
- A *buzzer* is used to create an alarm signal.
- A *meter* is used to measure quantities, such as in an ohmmeter, voltmeter, and ammeter.
- *Seven-segment displays* are used to indicate numbers and some letters.
- *Multidigit displays* can show several numbers at one time.
- A *piezoelectric crystal* vibrates when voltage is applied to it and is used to create frequencies.
- A *binding post* is a device for connecting wires temporarily.
- An *antenna* is a conductor used to detect electromagnetic radio waves and transfer the signal to a radio or television set.
- A *resistor* can be tested with an ohmmeter.
- A *capacitor* can be tested with an ohmmeter.
- An *inductor* can be tested with an ohmmeter.
- A *transformer* can be tested with an ohmmeter.

≡≡≡ SECTION 2–6
SELF-CHECKING QUIZ

Circle the most correct answer for each question.

1. Reading a resistor from left to right, the colors are orange, white, yellow, and gold. The value of the resistor is:

    a. 39.4 + 5%   b. 390 + 20%

    c. 3.9 k + 5%   d. 390 k + 5%

2. Reading a resistor from left to right, the colors are yellow, violet, silver, and silver. The value of the resistor is:

    a. 0.35 + 10%   b. 0.47 + 10%

    c. 4.7 + 20%   d. 35 + 10%

3. The size of a resistor depends on its:

    a. color code   b. current rating

    c. voltage rating   d. power rating

4. The volume control on a radio would be a:

    a. trimmer resistor b. potentiometer

    c. rheostat   d. none of the above

5. A device that changes resistance with a change of temperature is called a:

   a. zero resistor   b. photoresistor

   c. thermistor   d. potentiometer

6. The number 1000 is also referred to as the prefix:

   a. mega   b. micro

   c. kilo   d. pico

7. The prefix *pico* refers to the power-of-10 number:

   a. 10^{-12} b. 10^{-6}

   c. 10^3 d. 10^6

8. The number 345,000 can also be written as:

   a. 3.45×10^8 b. 345×10^4

   c. 345×10^3 d. none of the above

9. The unit of capacitance is the:

   a. ohm   b. farad

   c. henry   d. mho

10. The capacitor that must be connected in a circuit observing polarity is the:

    a. tubular type   b. disk type

    c. air type   d. electrolytic type

11. An inductor:

    a. opposes a change in current   b. is a coil of wire

    c. may have an iron core   d. all of the above

12. A transformer:

    a. can have more than one winding

    b. transforms the power in one winding to another

    c. can be a step-up or step-down type

    d. all of the above

13. A device used in one circuit to control another circuit is called a:

    a. relay   b. capacitor

    c. resistor   d. inductor

14. A device that converts sound energy into electrical energy is called a:

    a. motor   b. microphone

    c. loudspeaker   d. relay

15. A device used for measuring circuit conditions is the:

    a. loudspeaker   b. buzzer

    c. meter   d. none of the above

16. A seven-segment display can show:

    a. all numbers and letters

    b. words and special characters

    c. both of the above

    d. none of the above

17. A device that can generate a frequency when voltage is applied to it is the:

    a. motor   b. crystal

    c. capacitor   d. inductor

18. A device used to connect wires temporarily is the:

    a. resistor   b. relay

    c. binding post   d. buzzer

19. A device that converts electromagnetic energy into electrical energy is the:

    a. transformer   b. relay

    c. antenna   d. microphone

20. A device that will store a voltage charge temporarily is the:

    a. resistor   b. capacitor

    c. inductor   d. crystal

ANSWERS TO FILL-IN QUESTIONS AND SELF-CHECKING QUIZ

Experiment 1:   **(1)** zero   **(2)** range   **(3)** tolerance   **(4)** power or wattage

Experiment 2:   **(1)** charge   **(2)** discharge   **(3)** capacitors   **(4)** shorted
                      (5) open

Experiment 3:   **(1)** low   **(2)** shorted   **(3)** open

Self-Checking Quiz: **(1)** d   **(2)** b   **(3)** d   **(4)** b   **(5)** c   **(6)** c   **(7)** a   **(8)** c
                     (9) b   **(10)** d   **(11)** d   **(12)** d   **(13)** a   **(14)** b   **(15)** c
                     (16) d   **(17)** b   **(18)** c   **(19)** c   **(20)** b

Active Electronic Devices

INTRODUCTION
Active electronic devices are components that amplify or enhance the quality of a circuit. *Semiconductor* or *solid-state devices* as they are often called, operate on the principle of acting like a conductor under certain conditions and acting more like an insulator at other times with other conditions. Modern solid-state components are numerous and have various types of operations; however, the physical packages they come in are fairly standard and they are mounted on circuit boards using the same general methods.

UNIT OBJECTIVES
Upon completion of this unit, you will be able to:

1. Identify various types of semiconductor devices, such as diodes, bipolar transistors, field-effect transistors, thyristors, optoelectronic devices, and integrated circuits.
2. Draw the schematic symbols for the various solid-state devices.
3. Define the difference between standard electronic components and optoelectronic components.
4. Perform basic ohmmeter tests on the various devices.
5. Explain the reason for and use of heat sinks.
6. Describe the various types of solid-state packages.
7. Identify the schematic symbols of components in a schematic circuit drawing.
8. Identify the physical components on a circuit board.

≡ SECTION 3-1
FUNDAMENTAL CONCEPTS

3-1a DIODES

3-1a.1 Normal Diodes

Diodes are electronic devices that allow electrical current to flow in one direction but not in the other direction under certain conditions. Because they convert alternating current into direct current, they are also referred to as *rectifiers*. Diodes are absolutely necessary in a power supply, but are also found in nearly all types of electronic circuits. Figure 3-1 shows the schematic symbol for the diode and some physical packages. A diode has an anode lead at the flat end of the arrow and a cathode lead at the pointed end. The schematic symbol is painted on larger diodes to indicate the anode and cathode. On axial lead diodes, a band on one end indicates the cathode. It is important to connect a diode in a circuit in the proper direction.

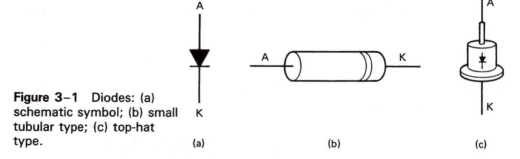

Figure 3-1 Diodes: (a) schematic symbol; (b) small tubular type; (c) top-hat type.

(a)   (b)   (c)

3-1a.2 Light-Emitting Diode

The *light-emitting diode* (LED) will glow when correctly connected in a circuit. LEDs can be found as pilot (on/off) lights on radios, TV sets, and other electronic equipment. An LED is shown in Figure 3–2. The notch or flat end of the package indicates the cathode.

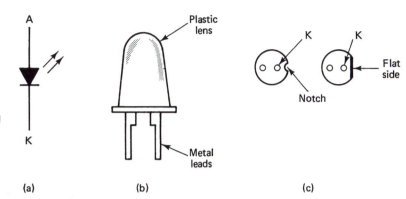

Figure 3–2 Light-emitting diode (LED): (a) schematic symbol; (b) physical construction; (c) lead identification (bottom view).

(a)   (b)   (c)

3–1a.3 Other Types of Diodes

Other types of diodes are used for specific functions. Schematic symbols for other types of diodes are shown in Figure 3–3 as an aid if you desire more information. Figure 3–4 shows a comparison of various diode packages. Larger packages can usually handle more current and dissipate more power.

Figure 3–3 Other types of diodes: (a) zener; (b) and (c) tunnel; (d) Schottky; (e) snap; (f) voltage-variable capacitor type.

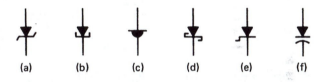

(a)   (b)   (c)   (d)   (e)   (f)

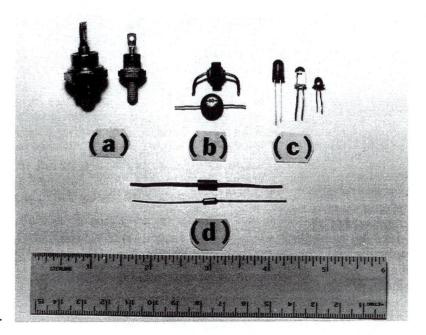

Figure 3–4 Diodes: (a) bolt type; (b) round type; (c) LEDs; (d) tubular type.

3-1a.4 Rectifier Modules

Some diodes are packaged as a bridge-type rectifier, as shown by the schematic symbol in Figure 3-5a. The packages will have a + and − terminal indicated on them.

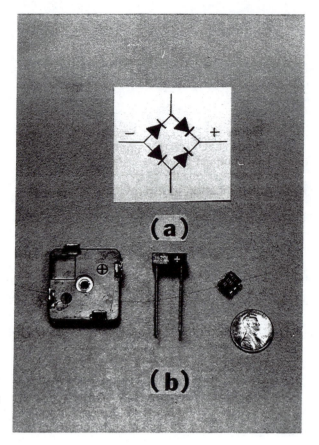

Figure 3-5 Rectifier modules: (a) schematic symbol; (b) types of packages.

3-1b THREE- AND FOUR-ELEMENT SEMICONDUCTOR DEVICES

3-1b.1 Bipolar Transistors

There are several types of transistors in use today. The older types are called *bipolar,* because current in the form of electrons and holes (the absence of electrons) flow through the device in opposite directions. Figure 3-6 shows the schematic symbols for bipolar transistors.

An *NPN transistor* has *n*-type semiconductor material for its collector (C), *p*-type material for the base (B), and *n*-type material for the emitter (E). The *PNP transistor* is the opposite, with its C having *p*-type, B having *n*-type, and E having *p*-type materials. The NPN and PNP bipolar transistors are used as electronically controlled switches and amplifiers. An *amplifier* is a circuit that increases a small input signal voltage for a more practical use, such as in a radio, TV, or stereo system.

A *unijunction transistor* (UJT) has *n*-type material going from base

Figure 3-6 Bipolar transistors: (a) NPN type; (b) PNP type; (c) UJT type.

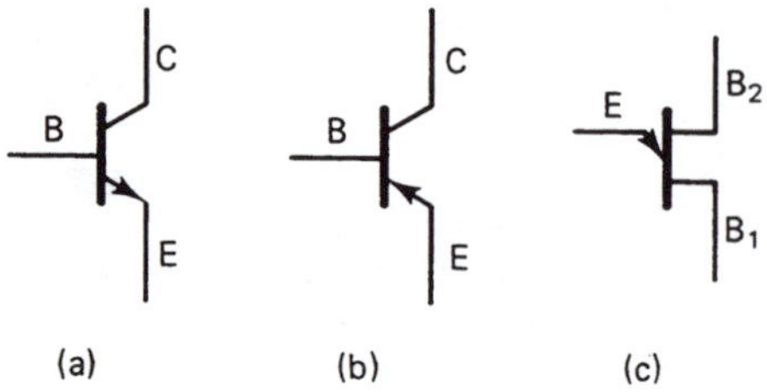

(a)   (b)   (c)

1 (B_1) to base 2 (B_2), with a small amount of *p*-type material midway, called the emitter (E). The UJT is normally used as a switch. Figure 3–7 shows some of the standard packages used for transistors and other semiconductor devices.

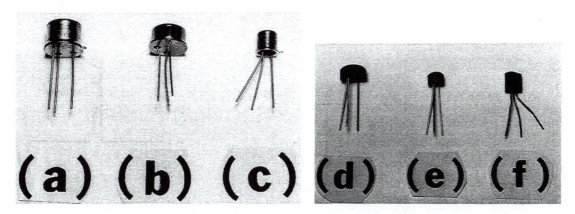

Figure 3–7 Semiconductor low-power packages: (a) TO-39 metal; (b) TO-5 metal; (c) TO-18 metal; (d) TO-5 plastic; (e) TO-18 plastic; (f) TO-92 plastic.

3–1b.2 Field-Effect Transistors

A *field-effect transistor* (FET) is a unipolar device in which current flows in one direction only. FETs are made of *n*- and *p*-type semiconductor material, but their three leads are called the drain (D), the gate (G), and the source (S). There are two general types of FETs: the *junction field-effect transistor* (JFET) and the *metal-oxide-semiconductor field-effect transistor* (MOSFET), of which several types are available, as shown in Figure 3–8.

Figure 3–8 Field-effect transistors: (a) *n*-channel JFET; (b) *p*-channel JFET; (c) *p*-channel depletion-type MOSFET; (d) *n*-channel depletion-type MOSFET; (e) *p*-channel enhancement-type MOSFET; (f) *n*-channel enhancement-type MOSFET.

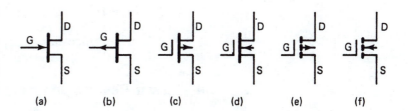

The MOSFET is very delicate to handle and can be destroyed by static voltage built up on the body or other objects. These devices are usually shipped with a shorting ring around their leads or pushed into antistatic foam, as shown in Figure 3–9a and b. Special care and anti-static equipment are used when working with MOSFET devices. Some MOSFETs have protection diodes built into them and short out voltages greater than ±15 V, as shown in Figure 3–9c and d.

3–1b.3 Thyristors

A *thyristor* is a multielement semiconductor device used for switching. Most thyristors are three-lead devices, as shown by the schematic symbols in Figure 3–10. These devices use the same types of packages as transistors.

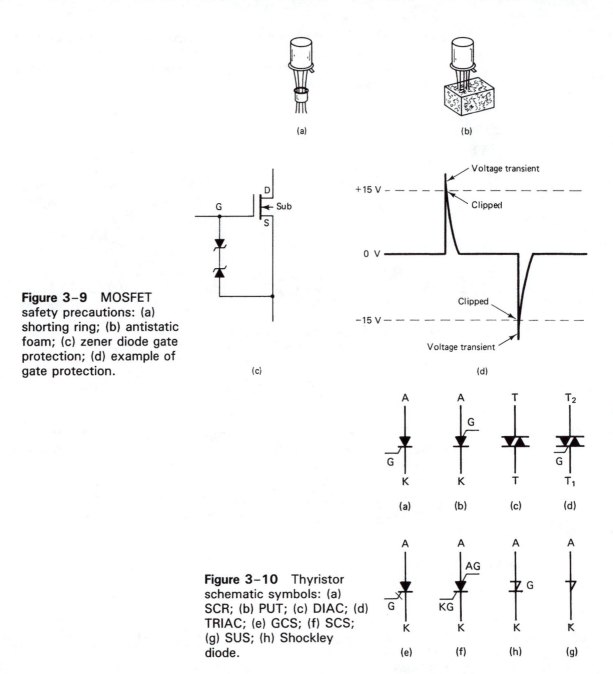

Figure 3-9 MOSFET safety precautions: (a) shorting ring; (b) antistatic foam; (c) zener diode gate protection; (d) example of gate protection.

Figure 3-10 Thyristor schematic symbols: (a) SCR; (b) PUT; (c) DIAC; (d) TRIAC; (e) GCS; (f) SCS; (g) SUS; (h) Shockley diode.

3-1b.4 Power Devices and Heat Sinks

Semiconductor components that operate motors, relays, loudspeakers, and other devices requiring large amounts of current are larger in size than the components mentioned thus far. These components are usually at the output of a circuit or system. The current that is in a circuit performs work, and the result is referred to as *power*. Components such as this are called power transistors, power FETs, and *power devices*. The power causes heat, that must be dissipated, for which the larger packages are more suitable. Figure 3–11 shows some packages used for power components.

Very often more heat must be removed from a component than the component is capable of doing. The component is then mounted on a device called a *heat sink*, which helps to dissipate heat into the surrounding atmosphere. Some types of heat sinks are shown in Figure 3–12. Packages TO-3 and TO-66 are mounted on single- and multiple-fin heat sinks (Figure 3–12a and b). The greater number of fins dissipates more heat.

Figure
3–11 Semiconductor power packages: (a) TO-3 metal high power; (b) TO-66 metal medium power; (c) TO-220 plastic; (d) TO-202 plastic tab medium power.

Figure 3–12 Heat sinks: (a) basic type for TO-3 packages; (b) fin-type for TO-3 packages; (c) push-on type for TO-5 packages.

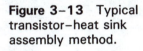

(a)   (b)   (c)

Some heat sinks are pushed over the smaller component devices (Figure 3–12c), such as TO-5 and TO-39 cases.

Figure 3–13 shows a typical heat sink assembly method. The two leads on the package must be insulated from the chassis with nylon

Figure 3–13 Typical transistor–heat sink assembly method.

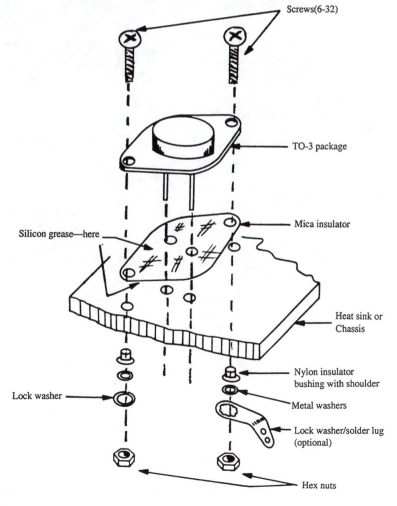

washers. The *chassis* is normally the metal part of a piece of equipment upon which electrical circuits are mounted. The body of the package is one of the three electrical leads and usually must be insulated from any metal parts. The *mica* insulates the body of the package from the chassis. *Silicon grease* is an insulator but provides additional heat transfer between the component and the heat sink.

Figure 3–14 shows some packages mounted on heat sinks. The larger power devices are mounted on fin-type heat sinks with nuts and bolts (Figure 3–14a–c). Smaller, low-voltage devices made of metal may have small heat sinks that push right over the top of the package (Figure 3–14d and e). Notice that even the medium-power TO-66 package can have a heat sink that pushes over its package (Figure 3–14f).

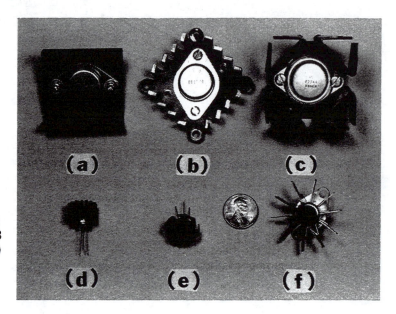

Figure 3–14 Semiconductor packages mounted on heat sinks; (a), (b), and (c) TO-3 on fin heat sink; (d) and (e) push-on heat sink for TO-39; (f) push-on heat sink for TO-66.

3–1c INTEGRATED CIRCUITS

An *integrated circuit* (IC) is a small device that may contain a few or many electronic circuits on a single base or foundation (also called a *substrate*). There are a few classifications used with ICs:

1. *Hybrid IC,* where two or more complete circuits are manufactured separately and then connected together on a common base.
2. *Monolithic IC,* where the complete circuits are composed of the same basic material and manufactured at the same time. This type of IC is the most used.
3. *Linear module (LM) IC,* which contains circuits that process voltage and current on a linear basis, such as an audio amplifier, voltage regulator, and timers. Some schematic symbols of linear devices are shown in Figure 3–15.
4. *Digital module (DM) IC,* where the circuits are designed as logic devices used in data switching arrangements. The most used basic logic devices, called *gates,* have various switching procedures. *Logic symbols* identify specific switching operations and are shown in Figure 3–16. *Flip-flops* are also digital circuits used for storing data temporarily in the form of voltage pulses. Logic symbols for some flip-flops are shown in Figure 3–17.

Other digital circuits include more complex devices such as microprocessors, memory, and small computers. Most of these digital devices

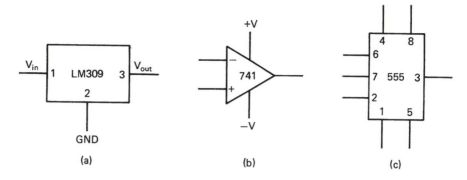

(a)   (b)   (c)

Figure 3–15 Linear integrated circuit schematic symbols: (a) voltage regulator; (b) op amp; (c) timer.

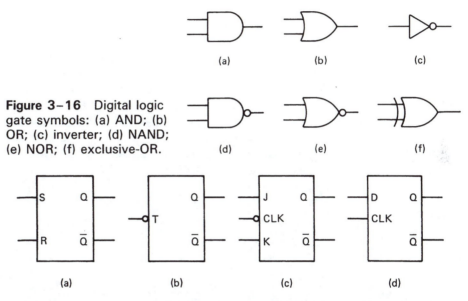

Figure 3–16 Digital logic gate symbols: (a) AND; (b) OR; (c) inverter; (d) NAND; (e) NOR; (f) exclusive-OR.

(a)   (b)   (c)   (d)

Figure 3–17 Digital flip-flop symbols: (a) *RS* flip-flop; (b) *T*-type flip-flop; (c) *JK* flip-flop; (d) *D*-type flip-flop.

are placed into dual-in-line (DIP) plastic or ceramic packages, as shown in Figure 3–18a. The connecting terminals or pins are arranged in a line along two sides of the package. They come in various sizes: 6-pin, 8-pin, 14-pin, 16-pin, 24-pin, up to 40-pin or larger. Some digital ICs are used in TO-5 and TO-66 packages, as shown Figure 3–18b.

Figure 3–18 IC packages: (a) dual-in-line (DIP) package; (b) ICs mounted in standard TO-5 and TO-66 packages.

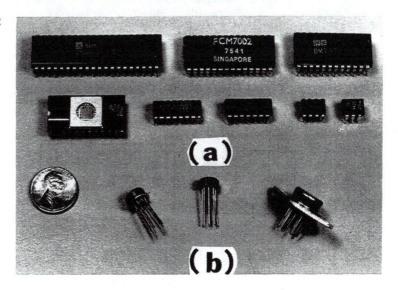

3–1d OPTOELECTRONIC DEVICES

Optoelectronics combines the technology of optics and electronics. These are devices that interact with light and electron current flow. A *photosource* produces light when current flows through it, such as an incandescent lamp or LED. A *photodetector* is sensitive to light and will allow current to flow when exposed to light.

3–1d.1 Photodetectors

Many solid-state devices are used as photodetectors, as shown by their schematic diagrams in Figure 3–19. These schematic symbols have an arrow pointing toward them, indicating that they are light sensitive. The components listed as photodiode, phototransistor, photodarlington, and photofet in Figure 3–19a–d, respectively, will allow more current to flow when exposed to light because their internal resistance has decreased. If the light is decreased, their resistance increases and less current flows through them. The light-activated silicon-controlled rectifier (LASCR) and photoTRIAC shown in Figure 3–19e and f act like latching switches when exposed to light. The photoresistor in Figure 3–19g acts similar to the phototransistor. The photovoltaic cell, often called a *solar cell*, shown in Figure 3–19h produces a small voltage and current when exposed to light.

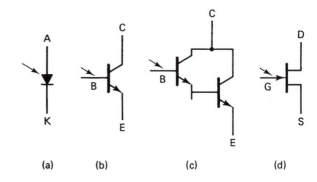

(a)   (b)   (c)   (d)

Figure 3–19 Optoelectronic schematic symbols:
(a) photodiode; (b) phototransistor; (c) photo darlington transistor; (d) photofet;
(e) LASCR; (f) photo TRIAC;
(g) photoresistor;
(h) photovoltaic cell.

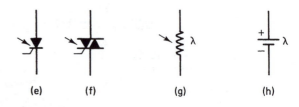

(e)   (f)   (g)   (h)

Figure 3–20 Optoisolator IC packages: (a) LED/phototransistor; (b) LED/photodarlington transistor; (c) LED/LASCR; (d) LED/photoTRIAC.

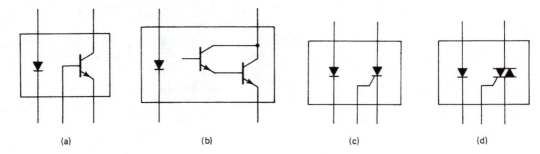

(a)   (b)   (c)   (d)

3–1d.2   Optoisolators

An *optoisolator* contains a photosource such as an LED and a photodetector within the same package, as shown in Figure 3–20. The type of photodetector used depends on the required operation of the circuit. Figure 3–21 shows some optoelectronic devices. The photodetector has a glass or plastic window in the package to allow light to enter. An optoisolator is usually mounted as a 6-pin DIP IC. Often, semiconductor components are not mounted directly in a circuit but are placed in sockets. Various sockets are shown in Figure 3–22.

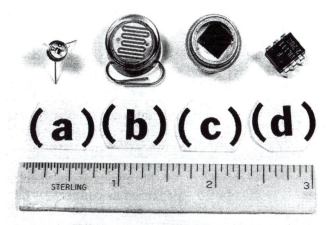

Figure 3–21 Optoelectronic devices: (a) phototransistor; (b) photoresistor; (c) photocell; (d) optoisolator.

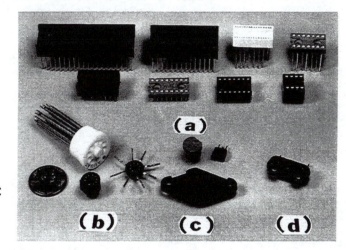

Figure 3–22 Semiconductor device sockets: (a) DIP IC types; (b) IC TO-5 types; (c) standard transistor types; (d) crystal type.

3–1e   THE SCHEMATIC DIAGRAM

All of the active and passive components that you have become familiar with are used in electronic circuits. Their schematic symbols may appear on a schematic diagram at one time or another. A *schematic diagram* is a complete electrical drawing of a circuit or group of circuits indicating the schematic symbols of electronic components. There are other symbols relating to the wiring connections on a schematic diagram which you should know, as shown in Figure 3–23.

Chassis ground (Figure 3–23a) is a common plane to which many circuits are connected. This is the structure upon which the circuits are built. The body of an automobile can be considered the chassis ground since it completes the electrical path for many circuits. The *earth ground* (Figure 3–23b) is a connection to planet earth, which supplies the bulk of electrons for our world. *Terminal points* (Figure 3–23c) are places in

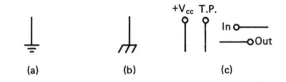

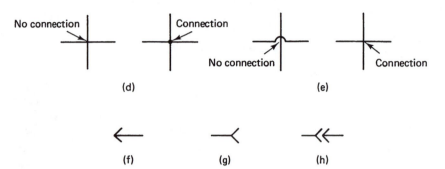

Figure 3–23 Ground and wire connections: (a) chassis ground; (b) earth ground; (c) terminal points; (d) wire connections with dot; (e) wire connections without dot; (f) male connector pin; (g) female connector socket; (h) connectors engaged.

a circuit where voltage is applied; input devices, output devices, and other circuits are connected; and places where testing can be carried out. Wire connections and nonconnections are shown in Figure 3–23d and e. The dot method is preferred.

Many circuit connections are used between circuit boards and are identified as connectors (Figure 3–23f–h). They are often referred to as *edge connectors* because they are found at the edge of circuit boards. The male connector pins are usually part of the board and fit into a female socket connector. Cables with both male and female connectors may be used.

Figure 3–24 Basic electronic circuit schematic diagrams: (a) power supply; (b) common-emitter transistor amplifier; (c) op amp; (d) digital logic diagrams.

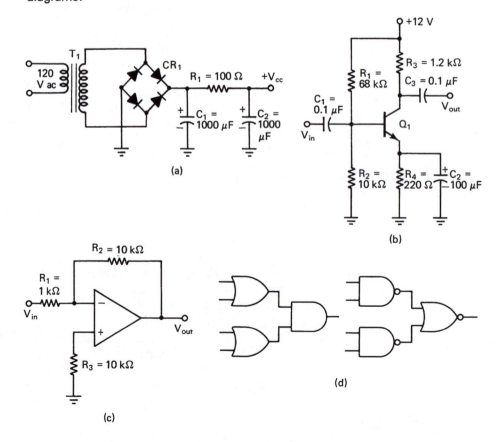

Some basic electronic circuits are shown in Figure 3–24. A power supply (Figure 3–24a) can be recognized by its transformer, diodes, or rectifier, and large capacitors. An amplifier (Figure 3–24b) will have an active device such as a bipolar transistor, JFET, or MOSFET, and several surrounding resistors and capacitors. An op amp circuit (Figure 3–24c) will have a minimum of components around it and can easily be identified by its triangular shape. Digital circuits (Figure 3–24d) are shown with their logic symbols.

≡ SECTION 3–2
DEFINITION EXERCISES

Write a brief description of each of the following terms.

1. Active device _____

_____

2. Diode _____

_____

3. LED _____

_____

4. Rectifier _____

_____

5. Bipolar _____

_____

6. NPN transistor _____

_____

7. PNP transistor _____

_____

8. Amplifier _____

_____

9. UJT _____

_____

10. JFET _____

_____

11. MOSFET _____

_____

12. Thyristor _____

_____

13. Power transistor _____

_____

14. Chassis _____

_____

15. Heat sink _____

_____

16. Mica _____

_____

17. Silicon grease _____

_____

18. Integrated circuit _____

_____

19. Substrate _____

_____

20. Hybrid IC _____

_____

21. Monolithic IC _____

_____

22. Linear module IC _____

_____

23. Digital module IC _____

_____

24. Gates _____

_____

25. Logic symbols _____

_____

26. Flip-flops _____

_____

27. DIP _____

_____

28. Optoelectronics _____

_____

29. Photosource _____

_____

30. Photodetector _____

_____

31. Solar cell _____

_____

32. Optoisolator _____

_____

33. Schematic diagram _____

_____

34. Chassis ground _____

_____

35. Earth ground _____

_____

36. Terminal points _____

_____

37. Edge connectors _____

_____

Complete this section before beginning the next section.

1. Draw the schematic symbols for the following components.

   a. diode   **b.** LED   **c.** zener diode   **d.** UJT

   e. NPN transistor   **f.** PNP transistor   **g.** *n*-channel JFET   **h.** *n*-channel depletion MOSFET

   i. *p*-channel enhancement MOSFET   **j.** SCR   **k.** TRIAC   **l.** op amp

2. Make a basic sketch of the following semiconductor packages.

   a. TO-5   **b.** TO-92   **c.** TO-220   **d.** TO-3

3. List the hardware parts required to mount a TO-3 package to a heat sink (refer to Figure 3–13).

   a.   **b.**

   c.   **d.**

   e.   **f.**

   g.   **h.** (optional)

4. Match the description in column A with the proper IC in column B.

Column A		*Column B*
____ **a.**	Two separate circuits connected together on the same base	**1.** LM
		2. monolithic
		3. DM
____ **b.**	All circuits made of same material	**4.** hybrid
____ **c.**	Circuits perform on a linear basis	
____ **d.**	Circuits use logic gates and flip-flops	

5. Draw the logic symbols for the following gates:

    a. AND   **b.** OR   **c.** inverter

    d. NAND   **e.** NOR   **f.** exclusive-OR

6. Draw the logic symbols for the following flip-flops.

    a. *RS*   **b.** *T*-type   **c.** *JK*   **d.** *D*-type

7. Draw the schematic symbol for the following optoelectronic devices.

    a. photodiode   **b.** phototransistor   **c.** photodarlington

    d. photofet   **e.** LASCR   **f.** photoTRIAC

    g. photoresistor   **h.** solar cell

8. Draw the following schematic circuit symbols.

    a. chassis ground   **b.** earth ground   **c.** terminal point

    c. crossed wires   **d.** wires connected   **e.** connectors
        (without dot)   (with dot)

9. Refer to Figure 3–25 and list the names of the devices in the schematic drawing that are numbered as below.

1.	2.	3.
4.	5.	6.
7.	8.	9.
10.	11.	12.

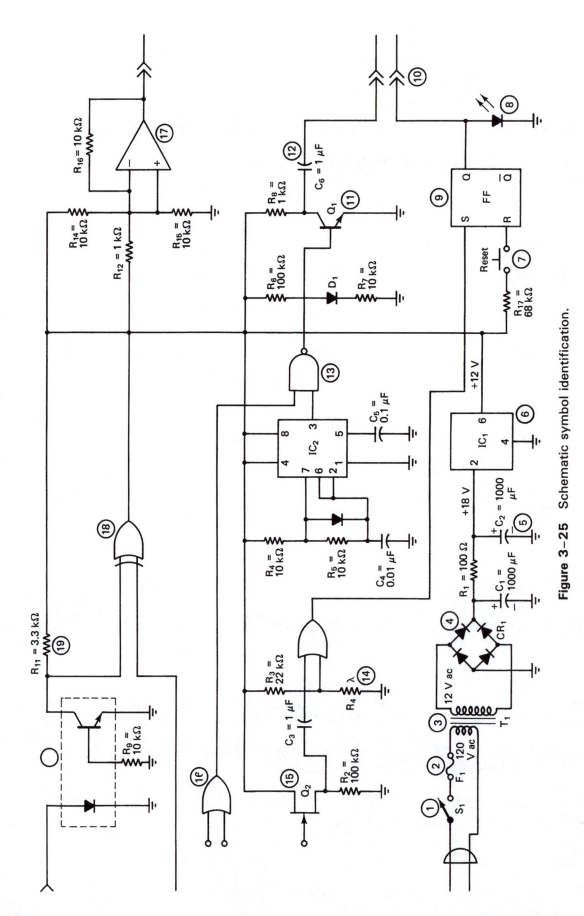

Figure 3–25 Schematic symbol identification.

13.　　　　　　　　14.　　　　　　　　15.

16.　　　　　　　　17.　　　　　　　　18.

19.　　　　　　　　20.

10. Refer to Figure 3–26 and identify and list the numbered components on the printed circuit board.

1.　　　　　　　　2.　　　　　　　　3.

4.　　　　　　　　5.　　　　　　　　6.

7.　　　　　　　　8.　　　　　　　　9.

10.　　　　　　　　11.　　　　　　　　12.

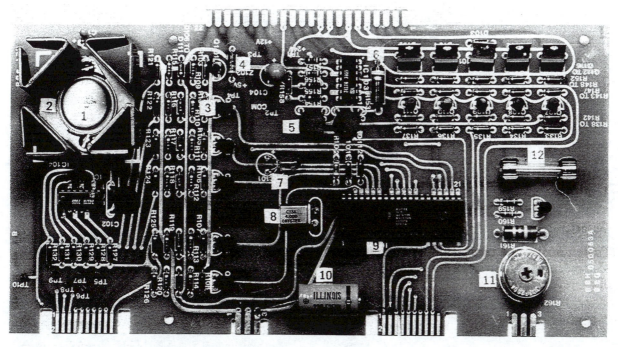

Figure 3–26　Component identification.

≣ SECTION 3–4
EXPERIMENTS

EXPERIMENT 1. Testing Semiconductor Diodes

Objective:

To demonstrate a practical method of testing diodes with an ohmmeter. This is called a *go/no go test.*

Introduction:

An ohmmeter has a low-voltage potential placed at its leads when measuring resistance. One lead is positive (usually red in color) and the other lead is negative (usually black in color). When the positive lead is placed on the anode of a diode and the negative lead on the cathode, the forward resistance (R_F) should be low, since the diode is forward biased. When the leads are reversed, the reverse resistance (R_R) should be high, since the diode is reverse biased. This simple go/no go test can determine if the diode is open or shorted.

Resistance measurements will vary with different types of diodes, but a high-to-low ratio of 100:1 is considered good for diodes. A shorted diode will show low-resistance readings in both directions and an open diode will show high resistance (infinity) in both directions.

Materials Needed:

A standard or digital ohmmeter
One or several diodes

Procedure:

1. Refer to Figure 3–27a and place the ohmmeter leads accordingly on the diode leads.
2. Set the ohmmeter to the lowest scale and record the R_F reading.
3. Refer to Figure 3–27b and place the ohmmeter leads accordingly on the diode leads.
4. Set the ohmmeter to the highest scale and record the R_R reading.

Figure 3–27 Testing a diode with an ohmmeter: (a) forward biased— minimum resistance (ideal = 0 Ω); (b) reverse biased—maximum resistance (ideal = ∞ Ω).

5. Calculate the ratio of reverse to forward resistance from the formula $R_R/R_F =$ ——— .

Fill-in Questions:

1. A forward-biased diode has _____

    _____ resistance.

2. A reverse-biased diode has _____

    _____ resistance.

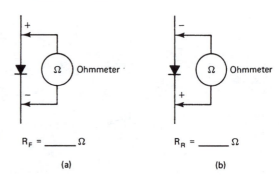

EXPERIMENT 2. Testing Bipolar Transistors

Objective:

To demonstrate a practical go/no go method of testing bipolar transistors with an ohmmeter.

Introduction:

For ohmmeter testing purposes, an NPN transistor is similar to two diodes back to back, as shown in Figure 3–28b. There exist two *pn* junctions, base–emitter and base–collector. When each of these junctions is forward biased by the ohmmeter—positive lead to *p* material and negative lead to *n* material (Figure 3–28c)—there should be a low-resistance indication. There should be a high-resistance reading when these junctions are reverse biased—positive lead to *n* material and negative lead to *p* material (Figure 3–28d).

The PNP transistor can be tested using a similar method, except that the diodes are face to face, as shown in FIgure 3–28e. This simple test determines if the transistor is shorted or open on a *go* (no problems)/*no go* (it has problems) basis.

Materials Needed:

A standard or digital ohmmeter

One or several bipolar transistors, including both types, NPN and PNP

Procedure:

1. Set the ohmmeter to the midrange scale.
2. Refer to Figure 3–28c to connect the ohmmeter to an NPN transistor for each junction and record the readings as high or low in the ohmmeter circles indicated.
3. Refer to Figure 3–28d to connect the ohmmeter to the NPN transistor for each junction and record the readings as high or low in the ohmmeter circles indicated.
4. Using a PNP transistor, perform the same procedure as in steps 1 to 3, while referring to Figure 3–28f and g.

Fill-in Questions

1. A forward-biased *pn* junction on a good

    bipolar transistor has _____ resistance.

2. A reverse-biased *pn* junction on a good

    bipolar transistor has _____ resistance.

3. A forward-biased *pn* junction with a high ohmmeter reading indicates that the

    transistor is _____ .

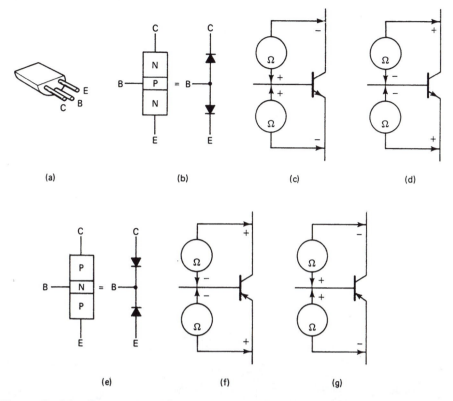

Figure 3-28 Testing bipolar transistors with an ohmmeter: (a) general package configuration; (b) NPN diode equivalent circuit; (c) and (d) NPN ohmmeter connections; (e) PNP diode equivalent circuit; (f) and (g) PNP ohmmeter connections.

4. A reverse-biased *pn* junction with a low ohmmeter reading indicates that the transistor is _____ .

EXPERIMENT 3. Testing JFETs

Objective:

To demonstrate a practical go/no go method of testing JFETs with an ohmmeter. This is a go/no go test.

Introduction:

For ohmmeter-testing purposes, the *n*-channel JFET is similar to a diode with its cathode connected to the middle of a resistor, as shown in Figure 3–29b. The ohmic resistance of the channel should be about the same regardless of the polarity of the ohmmeter lead connections from source to drain. With the positive lead on the gate, there should be a low-resistance reading when the negative lead is placed on the source or drain. The reading should be infinite when the negative lead is on the gate and the positive lead is placed on the source or drain. The same procedures are used for a *p*-channel

JFET, except that the diode's anode is connected to the resistor and the ohmmeter polarities are reversed.

Materials needed:

A standard or digital ohmmeter

One or several JFETs, including both *n*- and *p*-channel types

Procedure:

1. Set the ohmmeter to the midrange scale.
2. Refer to Figure 3–29c to connect the ohmmeter to an *n*-channel JFET and record the readings in the ohmmeter circles indicated.
3. Refer to Figure 3–29d to connect the ohmmeter to an *n*-channel JFET and record the readings in the ohmmeter circles indicated.
4. Using a *p*-channel JFET, perform the same procedures as in steps 1 to 3, while referring to Figure 3–29f and g.

Fill-in Questions:

1. For an *n*-channel JFET, with the positive lead on the gate and the

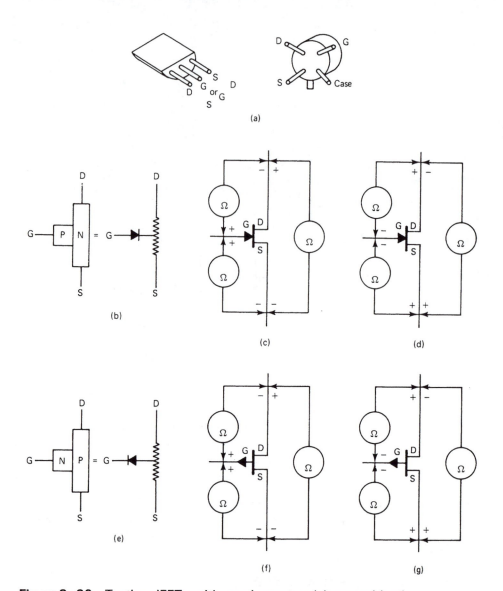

Figure 3-29 Testing JFETs with an ohmmeter: (a) general lead identification; (b) *n*-channel equivalent circuit; (c) and (d) *n*-channel ohmmeter connections; (e) *p*-channel equivalent circuit; (f) and (g) *p*-channel ohmmeter connections.

negative lead on the source, the ohmmeter should read _____ , compared to _____ or _____ when the leads are reversed.

2. For a *p*-channel JFET, with the positive lead on the gate and the negative lead on the drain, the ohmmeter should read

_____ or _____ , compared to _____ when the leads are reversed.

3. If the positive lead is placed on the drain and the negative lead is placed on the

source and the ohmmeter reads infinity, the JFET is _____ .

EXPERIMENT 4. Testing a UJT

Objective:

To demonstrate a practical go/no go method of testing a UJT with an ohmmeter. This is a go/no go test.

Introduction:

The *pn* junction of a UJT can be tested with an ohmmeter similar to testing diodes and bipolar transistors. With the negative lead placed on the emitter and the positive lead placed at either base, the junction is reverse biased and the resistance should be high or infinite. When the positive lead is placed on the emitter and the negative lead is placed at either base, the junction is forward biased and the resistance should be low. There should be a resistance reading of several thousand ohms when the meter is placed across the base leads.

Materials Needed:

A standard or digital ohmmeter
One or several UJTs

Procedure:

1. Set the ohmmeter to the midrange scale.
2. Refer to Figure 3–30a to connect the ohmmeter to the UJT properly for each lead and record the readings in the ohmmeter circles indicated.
3. Refer to Figure 3–30b to connect the ohmmeter to the UJT properly for each lead and record the readings in the ohmmeter circles indicated.

Fill-in Questions:

1. A forward-biased *pn* junction should have _____ resistance.

2. A reverse-biased *pn* junction should have _____ resistance.

3. A forward-biased *pn* junction with a high ohmmeter reading indicates that the UJT is _____ .

4. A reverse-biased *pn* junction with a low ohmmeter reading indicates that the UJT is _____ .

5. The resistance of a UJT from base to base reads the same regardless of the _____ of the ohmmeter leads.

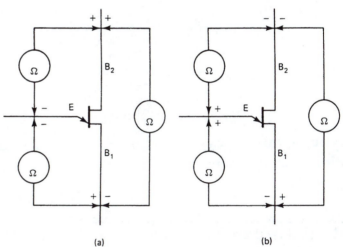

Figure 3–30 Testing a UJT with an ohmmeter.

(a)   (b)

EXPERIMENT 5. Testing an SCR with an Ohmmeter

Objective:

To demonstrate a practical go/no go method of testing an SCR with an ohmmeter. This is a go/no go test.

Introduction:

The PN junction from gate to cathode of an SCR can be tested with an ohmmeter similar to a regular diode. However, testing from anode to gate will not indicate if an SCR is working properly, because one of the PN junctions will always be reverse biased. The SCR can be tested with an ohmmeter by placing

the positive lead on the anode and the negative lead on the cathode with the gate left open. The meter should read high or infinite resistance. Placing a clip lead from the anode or positive lead of the ohmmeter to the gate will trigger the SCR, and the meter should indicate low resistance. When the clip lead is removed, the meter will continue to indicate low resistance if the power source is sufficient to produce the required holding current.

Materials Needed:

A standard or digital ohmmeter
One or several SCRs
A clip lead

Procedure:

1. Set the ohmmeter to the midrange scale.
2. Connect the ohmmeter to the SCR as shown in Figure 3–31b, and record the meter reading in the ohmmeter circle.
3. Connect the clip lead as shown in Figure 3–31c and record the reading of the meter in the ohmmeter circle.
4. Remove the clip lead as shown in Figure 3–31d and record the reading of the meter in the ohmmeter circle.

Fill-in Questions:

1. An SCR will have _____ resistance before being triggered.

2. An SCR will have _____ resistance after being triggered.

3. The _____-to-_____ resistance of an SCR can be checked like a normal diode.

4. An SCR is being tested with an ohmmeter. When the clip lead on the gate is removed, the meter indicates high resistance. This does not prove that the SCR is defective, but that the power source of the meter is not sufficient to

   produce the necessary _____

   _____ through the device.

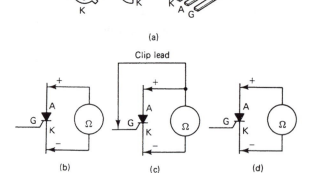

(a)

(b)   (c)   (d)

Figure 3–31 Testing an SCR with an ohmmeter: (a) general lead identification; (b) without clip lead; (c) with clip lead; (d) again without clip lead.

EXPERIMENT 6. Testing a TRIAC with an Ohmmeter

Objective:

To show how to test a TRIAC for conduction in both directions.

Introduction:

A TRIAC can be tested with an ohmmeter in a manner similar to testing an SCR or PUT. The positive lead of the ohmmeter is placed on T_2 and the negative lead is placed on T_1. The meter should read infinite resistance. A clip lead is placed from the positive lead to the gate, which should trigger on the TRIAC. The meter should now indicate low resistance. When the clip lead is removed, the meter will continue to indicate low resistance if the power source is sufficient to produce the required holding current. The meter leads are reversed on the main terminals of the TRIAC and a clip lead is placed from the negative lead to the gate to test for conduction in the reverse direction. This is a go/no go test.

Materials Needed:

A standard or digital voltmeter
One or several TRIACS
A clip lead

Procedure:

1. Set the ohmmeter to the low-range scale.
2. Connect the ohmmeter to the TRIAC as shown in Figure 3–32b and record the meter reading in the ohmmeter circle.
3. Connect the clip lead as shown in Figure 3–32c and record the meter in the ohmmeter circle.
4. Remove the clip lead as shown in Figure 3–32d and record the reading of the meter in the ohmmeter circle.
5. Connect the ohmmeter to the TRIAC as shown in Figure 3–32e and record the meter reading in the ohmmeter circle.
6. Connect the clip lead as shown in Figure 9–10f and record the reading of the meter in the ohmmeter circle.

7. Remove the clip lead as shown in Figure 9–10g and record the reading of the meter in the ohmmeter circle.

Fill-in Questions:

1. A TRIAC will have _____ resistance in either direction before being triggered.

2. A TRIAC will have _____ resistance in either direction after being triggered.

3. A TRIAC is being tested with an ohmmeter. When the clip lead is removed, the meter indicates high resistance. This does not prove that the TRIAC is defective, but that the power source of the meter is not sufficient to produce the necessary _____ _____ through the device.

4. If the ohmmeter shows low resistance before the TRIAC is triggered, this indicates that the TRIAC is _____ .

5. If the ohmmeter shows infinite resistance after the TRIAC is triggered, this indicates that the TRIAC is _____ _____ .

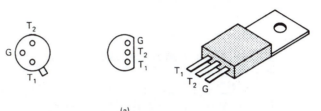

(a)

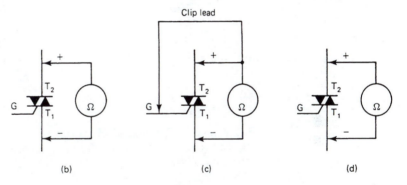

(b)   (c)   (d)

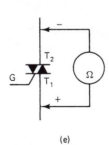

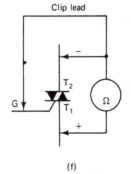

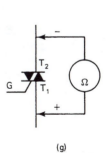

(e)   (f)   (g)

Figure 3–32 Testing a TRIAC with an ohmmeter: (a) general lead identification; (b) without clip lead; (c) with clip lead; (d) again without clip lead; (e) without clip lead; (f) with clip lead; (g) again without clip lead.

EXPERIMENT 7. Testing Photodetectors

Objective:

To show how to test photodetectors with a multimeter.

Introduction:

Most photodetectors can be tested with an ohmmeter by reading the resistance of the device when light is present and then when light is blocked. When light strikes the photodetector, the resistance should be low, and when light is blocked, the resistance should go high or infinite. Using the same procedure, the voltage of a solar cell can be tested with a voltmeter. This is a go/no go test.

Materials Needed:

A standard or digital multimeter

Assorted types of photodetectors

Procedure:

1. Testing a photodiode
   a. Connect the ohmmeter as shown in Figure 3–33a. Remember that the diode must be reverse biased.
   b. With light blocked from the diode, the ohmmeter reads _____ Ω.
   c. With light striking the diode, the ohmmeter reads _____ Ω.
2. Testing a phototransistor
   a. Connect the ohmmeter as shown in Figure 3–33b.
   b. With light blocked from the phototransistor, the ohmmeter reads _____ Ω.
   c. With light striking the phototransistor, the ohmmeter reads _____ Ω.
3. Testing a photodarlington transistor
   a. Connect the ohmmeter as shown in Figure 3–33c.

b. With light blocked from the transistor, the ohmmeter reads _____ Ω.
c. With light striking the transistor, the ohmmeter reads _____ Ω.
4. Testing a photofet
   a. Connect the ohmmeter as shown in Figure 3–33d, using a 1-MΩ resistor from gate to source.
   b. With light blocked from the photofet, the ohmmeter reads _____ Ω.
   c. With light striking the photofet, the ohmmeter reads _____ Ω.
5. Testing a LASCR
   a. Set the ohmmeter on the low range and connect it as shown in Figure 3–33e, using a 100-kΩ resistor from gate to cathode. To prevent false triggering, light must be blocked from the LASCR before and after the ohmmeter is connected.
   b. With light blocked from the LASCR, the ohmmeter reads _____ Ω.
   c. With light striking the LASCR, the ohmmeter reads _____ Ω.
   d. Again, with light blocked from the LASCR, the ohmmeter reads _____. Ω. (Is the ohmmeter supplying sufficient holding current to keep the LASCR on?)

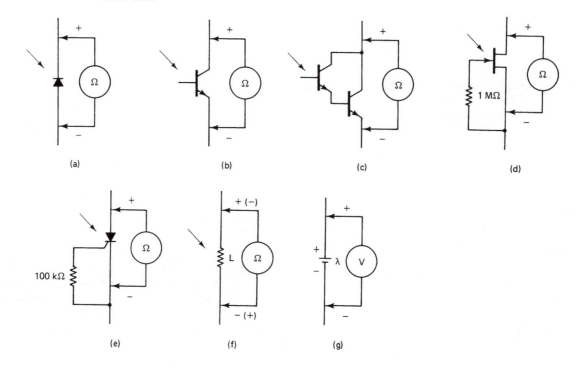

Figure 3–33 Testing photodetectors: (a) photodiode; (b) phototransistor; (c) photodarlington transistor; (d) photofet; (e) LASCR; (f) photoresistor; (g) solar cell.

Chap. 3 / Active Electronic Devices

6. Testing a photoresistor
   a. Connect the ohmmeter as shown in Figure 3–33f. The polarity of the leads is not critical.
   b. With light blocked from the photoresistor, the ohmmeter reads _____ Ω.
   c. With light striking the photoresistor, the ohmmeter reads _____ Ω.
7. Testing a solar cell
   a. Set the voltmeter on the low range and connect it as shown in Figure 3–33g.
   b. With light blocked from the solar cell, the voltmeter reads _____ V.
   c. With light striking the solar cell, the voltmeter reads _____ V.

Fill-in Questions:

1. When light is blocked from a photodetector, its resistance is _____ .

2. When light strikes a photodetector, its resistance is _____ .

3. When light strikes a solar cell, it produces _____ and _____ .

SECTION 3–5
INSTANT REVIEW

- An *active device* is a semiconductor device that increases or amplifies the voltage or current of a circuit.
- A *diode* allows current to flow in only one direction.
- An *LED* is a diode that emits light.
- A *rectifier* is another name for a diode.
- *Bipolar* means that the current flows in two directions simultaneously and is comprised of electrons and holes.
- An *NPN transistor* is a type of bipolar transistor.
- A *PNP transistor* is a type of bipolar transistor.
- An *amplifier* increases voltage or current in a circuit.
- A *UJT* is a semiconductor device used as a switch.
- A *JFET* is a unipolar device used for amplifying.
- A *MOSFET* is a unipolar device used for amplifying.
- *Thyristors* are semiconductor devices used for switching.
- A *power transistor* operates in a manner similar to a regular transistor, but can handle greater amounts of voltage and current.
- The *chassis* is the foundation on which all electronic circuits are built. It is usually made of metal.
- A *heat sink* is a metal device with fins that attaches to a semiconductor device to help dissipate heat.
- *Mica* is a thin, transparent insulating material used with heat sinks.
- *Silicon grease,* used on the mica and heat sink to help form a seal, also helps to transfer heat.
- An *integrated circuit* (IC) is a semiconductor device with several circuits contained in one small package.
- The *substrate* is the foundation upon which an IC is built.
- A *hybrid IC* contains separate circuits connected together in a single package.
- A *monolithic IC* has all of the circuits made from the same material within a single package.
- A *linear module IC* contains circuits that operate on a linear basis such as an audio amplifier.
- A *digital module IC* operates more as a switch.

- *Gates* is the term used for logic switching devices.
- A *logic symbol* is assigned to each gate, depending on its basic switching arrangement.
- *Flip-flops* are digital circuits used to store data temporarily.
- *DIP* stands for dual-in-line package.
- *Optoelectronics* is the combined technologies of optics and electronics.
- A *photosource* emits light. In electronics, a photosource converts electrical energy into light energy, such as an LED.
- A *photodetector* such as a photodiode or LASCR senses light energy, which changes its electrical characteristics.
- A *solar cell* is a photodetector that produces voltage and current when exposed to light.
- An *optoisolator* contains a photosource, such as an LED, and a photodetector, such as a phototransistor, in a single package.
- A *schematic diagram* is a complete electrical drawing showing schematic symbols.
- The *chassis ground* is the common plane to which many circuits are connected.
- The *earth ground* is a connection to the earth.
- *Terminal points* are places in a circuit where circuits or other devices are connected.
- *Wires joined by a dot* are electrically connected. Wires that cross but are not joined by a dot are not electrically connected.
- A *male connector pin* resembles an arrow.
- A *female connector pin* resembles the feather end of an arrow.
- When the *arrowhead* is placed into the feathered end, it indicates that the connectors are engaged.

≡ SECTION 3–6
SELF-CHECKING QUIZ

Circle the most correct answer for each question.

1. The component that is *not* an active device is the:

   a. NPN transistor   b. op amp

   c. UJT   d. resistor

2. A rectifier is also known as a:

   a. JFET   b. voltage regulator

   c. LED   d. diode

3. A circuit that increases voltage and current is:

   a. a voltage regulator   b. an amplifier

   c. a rectifier   d. a thyristor

4. Transistors that have current flow in both directions in the form of electrons and holes are called:

   a. hybrid   b. unipolar

   c. bipolar   d. none of the above

5. The component that *cannot* be used as an amplifying device is the:

   a. PNP transistor   b. JFET

   c. SCR   d. MOSFET

6. Heat sinks are used to:

   a. produce heat to keep transistors warm

   b. store heat energy for use at a later time

c. dissipate heat

d. all of the above

7. A device that requires careful handling because static electrical charges can damage it is the:

   a. UJT

   b. op amp

   c. LED

   d. MOSFET

8. The part of a piece of equipment upon which many circuits are constructed is called the:

   a. chassis

   b. substrate

   c. DIP

   d. none of the above

9. A material that is an electrical insulator is:

   a. copper

   b. mica

   c. *n*-type material

   d. all of the above

10. Silicon grease is used to:

    a. slide transistors into small places

    b. make conductors perform better

    c. provide a better heat transfer with a heat sink

    d. all of the above

11. A device that has many circuits made of the same material in a single package is the:

    a. chassis

    b. monolithic IC

    c. hybrid IC

    d. earth ground

12. Circuits performing switching operations are called:

    a. linear

    b. ICs

    c. digital

    d. none of the above

13. Circuits that can temporarily store data are called:

a. flip-flops

b. gates

c. op amps

d. LEDs

14. An LED is a:

    a. thyristor

    b. photosource

    c. photodetector

    d. none of the above

15. A photodetector device can be recognized by its:

    a. six leads

    b. flat-sided base

    c. black color

    d. window

16. Voltage is produced from the:

    a. phototransistor

    b. solar cell

    c. photofet

    d. photoTRIAC

17. A package containing an LED and a photo-resistor is called:

    a. a thyristor

    b. an op amp

    c. a timer

    d. an optoisolator

18. If a circuit required an amplifying device, it would use:

    a. an op amp

    b. a UJT

    c. a rectifier

    d. an AND gate

19. If a circuit were used to turn on a light, it would use:

    a. an SCR

    b. a UJT

    c. a photofet

    d. all of the above

20. If a piece of equipment needed a device to indicate that it was turned on, you would use:

    a. a diode

    b. an LED

    c. a phototransistor

    d. a solar cell

ANSWERS TO FILL-IN QUESTIONS AND SELF-CHECKING QUIZ

Experiment 1:   (1) low   (2) infinite

Experiment 2:   (1) low   (2) high   (3) open   (4) shorted

Experiment 3:   (1) low, high, infinite   (2) high, infinite, low   (3) open

Experiment 4:   **(1)** low   **(2)** high   **(3)** open   **(4)** shorted   **(5)** polarity

Experiment 5:   **(1)** high   **(2)** low   **(3)** gate, cathode   **(4)** holding current

Experiment 6:   **(1)** high   **(2)** low   **(3)** holding current   **(4)** shorted
                         (5) open

Experiment 7:   **(1)** high   **(2)** low   **(3)** voltage, current

Self-Checking Quiz: **(1)** d   **(2)** d   **(3)** b   **(4)** c   **(5)** c   **(6)** c   **(7)** d   **(8)** a
                           (9) b   **(10)** c   **(11)** b   **(12)** c   **(13)** a   **(14)** b   **(15)** d
                           (16) b   **(17)** d   **(18)** a   **(19)** d   **(20)** b

Unit 4

Basic Tools and Their Use

INTRODUCTION The assembly of electronic equipment requires the knowledge and skills of basic hand tools, power tools, and soldering tools. In performing their tasks, both the electronics assembler and the electronics technician must handle hardware. To assemble and disassemble mechanical parts and hardware, you need to know how to use basic hand tools and power tools correctly.

UNIT OBJECTIVES Upon completion of this unit, you will be able to:

1. Identify basic hand tools.
2. Explain the correct use of hand tools.
3. Show safety procedures for using hand tools.
4. Perform the correct steps for stripping insulated wire.
5. Give reasons for using a vise.
6. Explain the use of clamps.
7. Demonstrate how to use an electric hand drill.
8. Use an electric jigsaw correctly.
9. Explain how to use a drill press.
10. Explain how to use a wheel grinder.
11. List safety precautions when working with power tools.
12. Describe safety devices and their use.
13. Select the proper tools for specific jobs.
14. Determine good-quality tools that last longer and are more efficient to use.
15. Repair tools or effect a replacement.

FUNDAMENTAL CONCEPTS

4–1a BASIC HAND TOOLS

Nearly all the mechanical work in electronics involves the use of metal or plastics; therefore, only the appropriate tools will be mentioned. Figure 4–1 shows a few basic hand tools.

4–1a.1 Tool Description

The *hacksaw* (Figure 4–1a) was originally used for cutting metal pipes and some types of sheet metal. It has a special blade of hardened steel

Figure 4–1 Basic tools: (a) hacksaw; (b) claw hammer; (c) ball peen hammer; (d) slip-joint pliers; (e) channel-lock pliers; (f) vice-grip pliers.

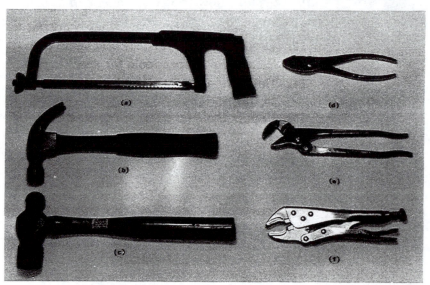

that can be used several times for cutting steel before becoming dull and needing replacement. The hacksaw can also be used on some types of plastic.

The *nail* or *claw hammer* (Figure 4–1b) has a steel head and a wooden or fiberglass handle and is used to drive nails into wood and pull nails out of wood. The claw part of the hammer can also be used to pry metal components apart.

Safety Note! Take your time when striking a nail; position the hammer in the center of the striking surface.

The *ball peen hammer* (Figure 4–1c) has a metal head and wood or fiberglass handle and is used to drive chisels and punches. It can also be used to shape and straighten unhardened metal.

Safety Note! Strike the hammer squarely and avoid glancing blows that may cause the edge of the face of the head to chip, possibly resulting in eye or other serious injury. Never strike one hammer against another.

Slip-joint pliers (Figure 4–1d) are made of steel and have a set of gripping jaws with two handles. An elongated slot where the jaws are held together by a heavy pin allows the jaws to slide to a wider holding position. Pliers are used for gripping, turning, and bending.

Channel-lock pliers (Figure 4–1e) are similar to slip-joint pliers except that they have more elongated slots for various positioning of the jaws. They are better suited for gripping round, square, flat, and hexagonal objects. Channel-lock pliers are capable of applying limited torque (circular motion) without damage to the workpiece.

Vise-grip pliers (Figure 4–1f) are a combination tool that functions as pliers, a wrench, and a limited-pressure portable vice or clamp. The screw at the end of one handle is turned to adjust the width of the jaws to a little less than the size of the object to be held. The pliers are then opened and placed on the object. When the handles are closed a locking action takes place which clamps the pliers to the object.

Safety Note! Never use pliers as a hammer or strike them with a hammer or other tool. Use larger pliers for tough jobs.

Figure 4–2 shows some workpiece holding devices. The *bench vise* (Figure 4–2a) is an extremely important device for holding workpieces and providing safety for cutting, drilling, and assembling parts.

Safety Note! Make sure that the vise is securely mounted to the

Figure 4–2 Clamping devices: (a) vise; (b) C-clamps.

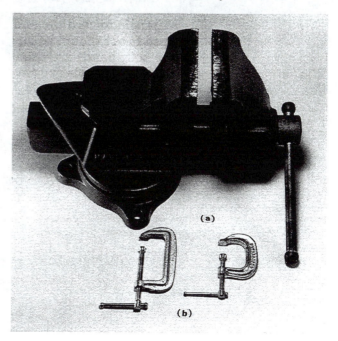

(a)

(b)

bench, using all of the holes provided in the base of the vise. Replace or discard any worn or loose parts, such as the turning handle or moving jaw. Discard any vise with hairline cracks. Replace worn jaw inserts. Lightly oil all moving parts.

When using a vise:

- Never tighten the handle beyond hand pressure.
- Never use an extension handle for extra clamping pressure.
- Never pound on the handle to tighten beyond hand pressure.
- Avoid clamping the workpiece with heavy pressure at the corner of the vise jaws, as this may break off a corner of a jaw.
- When clamping an extra-long workpiece, support the far end of the workpiece rather than putting extra pressure on the vise.
- When a workpiece is held in the vise for sawing, saw as close to the jaws as possible, to prevent vibration. Be careful not to cut the jaws.
- Use jaw liners with a vise if there is any possibility of marring the workpiece.
- Never pound or use the jaws of the vise as an anvil.

Another very useful holding device is the *C-clamp* (Figure 4–2b), which resembles the letter "C" and has a threaded shaft or spindle connected to a clamping surface. These clamps are used to hold small parts together for gluing, soldering, or assembling.

Safety Note! Discard any C-clamp that has a bent frame or bent spindle. Do not tighten the spindle beyond hand pressure. Before using, make sure that the swivel at the end of the threaded shaft turns freely. Lightly oil the moving parts, but make sure that other parts coming in contact with the workpiece are clean. Use pads with C-clamps to avoid marring a workpiece.

4–1a.2 Using the Hacksaw

When using the hacksaw, the first consideration is to make sure that the blade is sharp and mounted in the proper direction. The saw teeth of the blade point more toward one direction than the other. If you run your finger lightly over the blade, you will feel one direction smoother than the other. A properly mounted blade has the teeth pointing from the handle toward the end of the brace of the saw. The workpiece you intend to cut should be placed into a vice as shown in Figure 4–3. Very often both hands are used to operate the saw, to achieve an even and straight cut. Initially, you may want to place the saw on the workpiece

Figure 4–3 Using a hacksaw.

closer to the handle and then pull or draw the saw back toward you to obtain a starting grove that makes the sawing easier.

Safety Note! Try to use a sharp blade at all times. Make sure that the blade is tightly mounted in the saw. Be careful, and take your time when sawing. Try not to let the saw slip out of the grove or line it is cutting. A saw that slips may damage the workpiece or injure the hand.

4–1a.3 Using Pliers

Pliers can be used for holding nuts, as shown in Figure 4–4. They can also be used to loosen nuts and bolts, but care should be used not to slip and burr or round off the nut, which will make getting it off much more difficult. Pliers are also used for bending light metal.

Figure 4–4 Using slip-joint pliers.

Safety Note! Always use pliers on moderate holding jobs. Other tools are better to use on difficult nuts and bolts. Never use the pliers as a hammer and do not hit the pliers with a hammer. Be careful not to pinch your hand when using the pliers.

4–1a.4 Using Wrenches

Wrenches are tools for tightening and loosening nuts and bolts. They are better suited and safer to use than pliers when tightening and loosening nuts. The rigid part that fits over or around the nut is less likely to slip and cause damage to the nut or person using the tool. Various wrenches are used to tighten and loosen nut and bolts. Figure 4–5 shows an assortment of wrenches.

An *open-end wrench* has an opening in the part that fits over a nut. When using an open-end wrench (Figure 4–5a), place the tool on the nut with the larger curved portion on the nut so that it has a pulling action as shown in Figure 4–6 when the wrench is pulled in the direction shown by the arrow.

A *box-end wrench* has a completely closed part that fits over the nut. A box-end wrench (Figure 4–5b) is better to use on a nut that is "frozen" and needs a lot of pressure to loosen it. Combination open-end and box-end wrenches (Figure 4–5c) are used in conjunction with removing a nut. First, the box-end wrench is used to loosen the nut and the open-end wrench is used to remove the nut from the bolt much faster.

An *adjustable wrench* (Figure 4–5d) has a movable part that fits over the nut and can be adjusted to fit nuts of many sizes. An adjustable or *crescent wrench,* as it is sometimes called (Figure 4–5d), is very useful when removing or tightening various-sized nuts on the same piece of equipment.

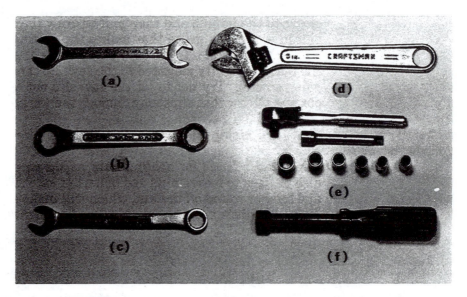

Figure 4–5 Wrenches: (a) open-end; (b) box-end; (c) combination open-end, box-end; (d) adjustable; (e) rachet wrench with extension and sockets; (f) nut driver.

Figure 4–6 Using an open-end wrench.

A *ratchet wrench* (Figure 4–5e, often called a *socket wrench*) has a handle with a ratchet gear, which enables it to move a slight distance to tighten a nut and then can be pushed in the opposite direction without loosening the nut. This feature allows the ratchet wrench to reach places where a normal wrench does not have the room to turn. Extensions and various-sized sockets fit onto the ratchet wrench, making it a versatile and important tool for many applications.

A *nut driver* (Figure 4–5f) has a handle like a screwdriver but has a hex socket on the end of the shaft. The sockets come in various sizes, and the use of this tool greatly speeds up assemble and disassemble time. This tool is particularly well suited for holding nuts while the screw is tightened from the opposite side.

Safety Note! Do not use a wrench as a hammer. Do not pound on a wrench with a hammer to loosen a nut, since this may distort the flat side of the face of the wrench. Never use an extension on the wrench handle, such as a pipe, because this could bend or damage the tool and might slip and cause injury to the person using the tool. Do not use an open end or adjustable wrench on a "frozen" nut. Use a box-end wrench on a "frozen" nut. Always check wrench openings for wear. A worn wrench may slip on a nut, causing damage to the nut or person using the tool.

4–1a.5 Using Screwdrivers

There are many types of screwdrivers used in industry today, but the two main types are shown in Figure 4–7. The *blade* or *standard screwdriver* (Figure 4–7a) has a flat blade that fits into screws and bolts for turning. The *Phillips-head screwdriver* (Figure 4–7b) has four edges similar to a cross and fits into Phillips-type screw heads for turning.

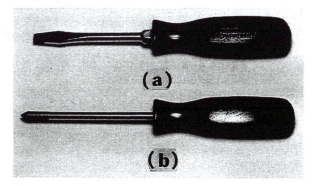

Figure 4–7 Screwdrivers:
(a) flat-blade type; (b)
Phillips-head type.

Safety Note!   You should never use a screwdriver on a small object held in your hand as shown in Figure 4–8a. The screwdriver may slip and puncture your hand. Small objects that require the use of a screwdriver should be put into a vice as shown in Figure 4–8b, if at all possible. Objects that cannot be put into a vice should be placed on a workbench or some other solid base and be held firmly while using a screwdriver. Never use the screwdriver as a chisel.

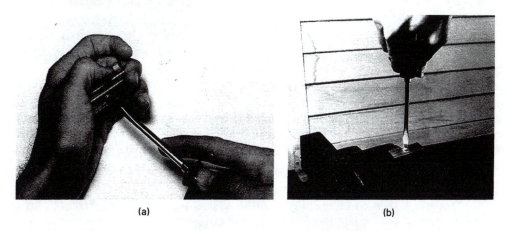

(a)   (b)

Figure 4–8 Using a screwdriver: (a) incorrect; (b) correct.

More basic hand tools are shown in Figure 4–9.

4–1a.6 Hex Wrench

The *hex* or *Allen wrench* (Figure 4–9a) is a small six-sided tool with a 90° bend. The hex wrench is used on recessed screws that hold gears, pulleys, and knobs on shafts as shown in Figure 4–10. The *spline* or *Bristo wrench* looks like the hex wrench but has a different type of shaft structure.
   Safety Note!   Never use a worn hex wrench, because it may slip and strip the screw head. Make sure that the wrench fits snuggly into the screw head.

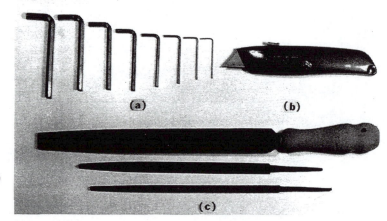

Figure 4–9 More tools: (a) Allen wrenches; (b) utility cutter; (c) files.

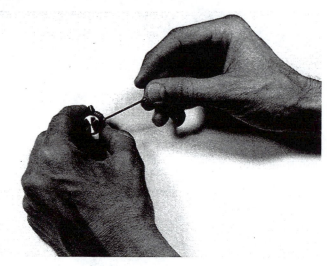

Figure 4–10 Using an Allen wrench.

4–1a.7 Utility Cutter

The *utility cutter* (Figure 4–9b), sometimes referred to as a carpet knife, has replaceable blades similar to a safety razor which fits into the handle. The cutter may have a lever for retracting the blade back into the handle when it is not in use. New blades can be stored in the handle. It is very important to use extreme care when using the cutter. Parts of the body should not be placed in the path of the cutter as shown in Figure 4–11a. If the cutter should slip, a very serious cut could result. Figure 4–11b shows the correct way to place the hand out of the path of the cutter.

Figure 4–11 Using a utility cutter: (a) incorrect; (b) correct.

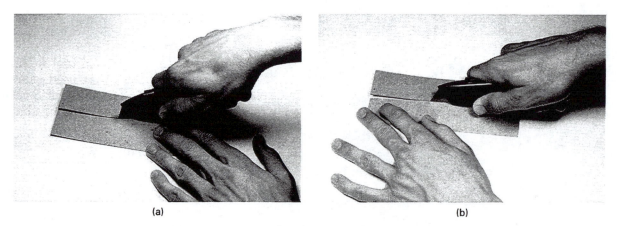

(a)   (b)

Safety Note! Use only a sharp blade in the cutter. Take your time when using the cutter, and make steady, even cuts.

4–1a.8 Metal Files

Metal files (Figure 4–9c) are abrasive devices used to remove small amounts of metal from parts by abrasive action. There are many types and sizes of metal files. The cross-sectional shape of a file may be flat, round, half-round, and triangular. Place the workpiece to be filed into a vice as shown in Figure 4–12. Most of the time, two hands can control the amount of filing to be done. Metal files are mostly used to remove sharp and dangerous sharp burrs from cut metal parts.

Figure 4–12 Using a file.

Safety Note! Metal files used on steel or iron usually have no problem, but using a file on aluminum will clog the fine abrasive parts, which may cause the file to slip and cause damage. A small, stiff, wire brush can be used to remove the aluminum from the file so that it can be used again.

4–1a.9 Electricians's 6-in-1 Tool

The *electrician's 6-in-1 tool* shown in Figure 4–13a is basically a crimping tool used on the wire connectors shown in Figure 4–13b. However, this tool is much more than just a crimping tool, as shown in Figure 4–14a. The electrician's 6-in-1 tool can be used to cut wire (Figure 4–14b), to crimp insulated terminals (Figure 4–14c), to crimp noninsulated terminals (Figure 4–14d), as a thread gage for checking the size of some screws

Figure 4–13 Electrician's 6-in-1 tool: (a) tool; (b) wire crimp splice and various crimp-type terminals.

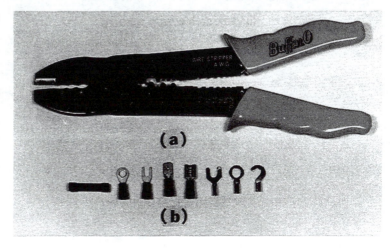

and bolts (Figure 4–14e), as a bolt cutter that uses the same part as the thread gages (Figure 4–14f), and as a wire stripper (Figure 4–14g).

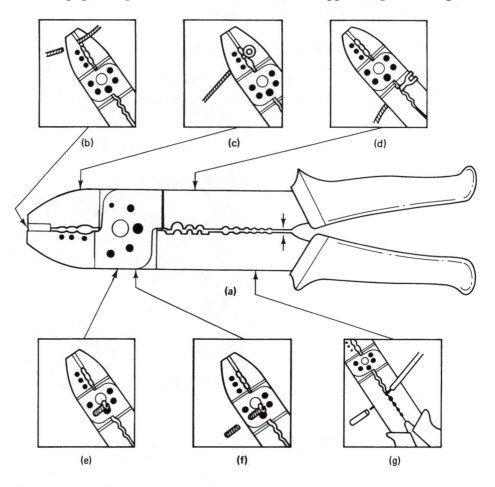

Figure 4–14 Electrician's 6-in-1 tool: (a) tool; (b) cutting wire; (c) crimping insulated terminals; (d) crimping nonsinulated terminals; (e) thread gages; (f) bolt cutter; (g) wire stripper.

4–1b ELECTRONIC HAND TOOLS

Basic electronic hand tools are shown in Figure 4–15. These tools are used for preparing wires and components for assembly and for replacing wires and components during a repair job.

Figure 4–15 Wire and electronic component tools: (a) adjustable wire stripper; (b) diagonal cutters; (c) long-nose pliers.

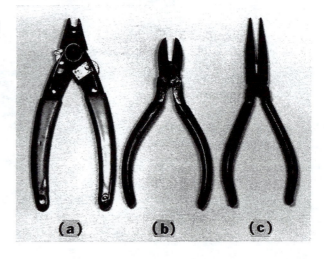

4–1b.1   Wire Stripper

The *wire stripper* (Figure 4–15a) is a tool that can be adjusted to strip the insulation off wire without cutting or nicking the strands of wire. The size of the hole in the stripper that allows the copper wire to pass is adjusted by a screw on the stripper, as shown in Figure 4–16a. The screw slides back and forth, as shown by the double-headed arrow, to set the size of the hole, and then it is tightened to hold that position. The stripper is placed over the wire as shown in Figure 4–16b. The handles are squeezed tightly, which cuts through the insulation and the stripper is pulled off the wire.

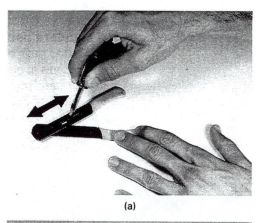

(a)

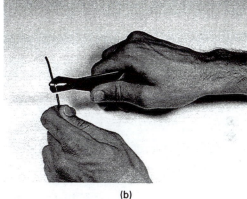

Figure 4–16   Using wire strippers: (a) adjusting wire size; (b) stripping wire.

(b)

Safety Note!   Keep the pivot pin lightly oiled. Do not use a wire stripper that is worn or loose. Pull gently when stripping wire and aim the stripping end of the wire away from you.

4–1b.2   Diagonal Cutters

Diagonal cutters, often referred to as "dikes" (Figure 4–15b), resemble a pair of pliers, but the tool end has cutting edges instead of grippers. Diagonal cutters are used for cutting wire and in some cases, stripping wire. Figure 4–17 shows how diagonal cutters are used to cut a wire.

Safety Note!   The end of the wire to be cut should be pointed downward and away from you. The force of the cutting edges can cause the small cut wire to become a projectile that can hit someone in the face or eye and cause serious damage. Do not cut anything else except copper wire and electronic component leads. Very hard materials, such as steel, can nick the cutting edge of the diagonal cutters, which makes them unable to cut copper wire properly.

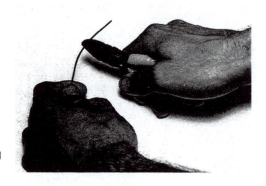

Figure 4-17 Using diagonal cutters.

4-1b.3 Long Nose Pliers

A standard pair of *long nose pliers* (Figure 4–15c) is shaped similar to a regular pair of pliers, except that the jaws are long and pointed. Long nose pliers are used for holding components, as a heat sink, and for shaping or bending wires and component leads, as shown in Figure 4–18.

Safety Note! Keep the movable pin of the pliers lightly oiled. Be careful not to break the tip of one of the jaws. Do not use too much pressure when holding an object, which might cause the tips to break. This problem will make it difficult to hold other components.

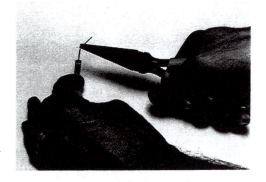

Figure 4-18 Using long-nose pliers.

4-1c BASIC ELECTRIC TOOLS

Many electronic assembly jobs require the use of electric tools for fast and efficient production.

4-1c.1 The Hand Drill

The *hand drill* is a motorized device used primarily for drilling holes. Other attachments are available for the hand drill so that it can be used as a screwdriver, socket wrench, sander, wire brush, and buffer. Figure 4–19 shows a hand drill and some of its drill bits.

The drill (Figure 4–19a) has a motor and a chuck that holds the drill bits (Figure 4–19c) used for drilling holes. Various-sized chucks will accommodate drill bits up to $\frac{1}{4}$, $\frac{3}{8}$, and $\frac{1}{2}$ in. The chuck key (Figure 4–19b) is used to tighten the drill bit in the chuck, as shown in Figure 4–20. A hole cutter (Figure 4–19d) is a regular drill bit with a larger circular saw-type blade for cutting larger holes than a drill bit can produce.

Before drilling metal, it is necessary to place a small starting hole at the place to be drilled. A center punch and hammer is used to make the starting hole, as shown in Figure 4–21.

Very often, with small workpieces, the drill bit will dig into the metal and it will spin rapidly, causing it to be like a rotating knife or blade,

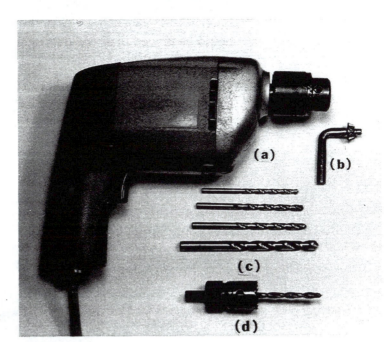

Figure 4–19 Electric hand drill: (a) tool; (b) chuck key; (c) drill bits; (d) hole cutter.

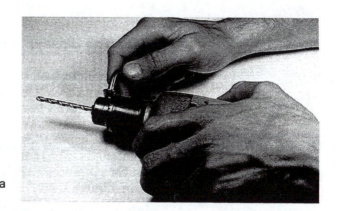

Figure 4–20 Tightening a drill bit with a chuck key.

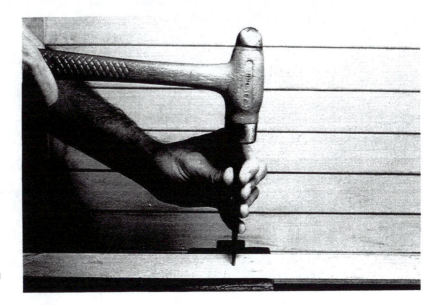

Figure 4–21 Using a center punch.

which can cause severe cuts to the hand. Small workpieces to be drilled should be placed in a vice during the drilling, as shown in Figure 4–22.

Safety Note! Check the drill for frayed wires and make any repairs where needed. Make sure that the bit is mounted in the chuck properly. Use the chuck key to tighten the drill bit in the chuck. Be careful not to get any flying chips in your eyes. Wear safety goggles or glasses if possible. Place small workpieces in a vice when drilling.

Figure 4–22 Drilling material that is well secured.

4–1c.2 The Jigsaw

A *jigsaw* is a motorized saw with changeable blades as shown in Figure 4–23. The saw is used to cut thin pieces of material, such as metal, plastic, and wood. It is important for safety reasons to place the blade into the blade holder correctly. The blade fits into a slot and then a screw is tightened to hold it in place. Special care should be used when using the jigsaw to cut materials. The material being cut must be held securely while being cut. Small pieces being cut by a jigsaw must be placed in a vice as shown in Figure 4–24.

Safety Note! Check the jigsaw for frayed wires and make any repairs where needed. Make sure that the blade is properly installed before sawing. Hold the workpiece securely and watch out for any flying bits of material. Place small workpieces in a vice before sawing.

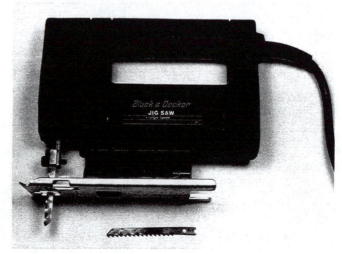

Figure 4–23 Jigsaw and blades.

Figure 4–24 Using a jigsaw.

4–1c.3 The Drill Press

A *drill press* is a large motorized drill mounted in a vertical position on a special stand as shown in Figure 4–25. The drill bits are placed into the chuck the same as a hand drill, and a chuck key is used for tightening. This tool is a must for precision drilling and speeds up the time required to drill many holes at one time.

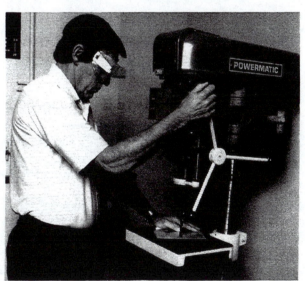

Figure 4–25 Using a drill press. (Note safety glasses.)

Safety Note! Wear safety goggles or glasses when using a drill press. Be careful not to let any of your clothing get caught in the drill bit. In such a situation, the off switch is not always easy to reach. Even if the power is turned off, the large motor still requires some time to come to a complete stop and could cause considerable damage. Read all safety signs attached to the drill. Small workpieces should be placed into a special drill vice.

4–1c.4 The Bench Grinder

A *bench grinder* is a motor with one or two abrasive stone wheels connected on the end. The grinder is used to smooth off rough edges of materials, to sharpen tools such as drill bits and screwdrivers, and to trim workpieces for proper fitting in an assembly. The grinder has a small table in front of the wheel where the material to be ground is placed as shown in Figure 4–26.

Safety Note! Wear safety goggles when using the grinder to prevent chips of material from hitting the eyes. Be careful not to get your hand or any part of the body against the spinning wheel. Hold the workpiece securely against the wheel, but do not overforce it to cause splintering or other problems.

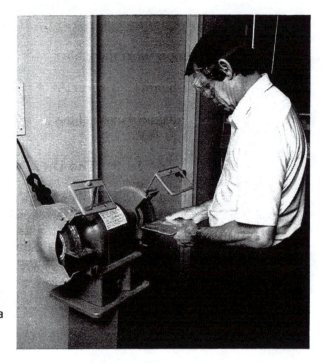

Figure 4–26 Using a grinder. (Note safety glasses.)

4–1c.5 Sheet-Metal Tools for Chassis Fabrication

Normally, electronic assemblers do not produce metal chassis, but it may be part of the process in an equipment manufacturing facility. The metal received at the factory will be in sheets. The sheets will be cut, holes or openings punched or cut out, then the sheets drilled for screws and bolts and formed (or bent) into the required shape of the chassis. The tools used to prepare sheet metal may be manually operated or powered by pneumatic, hydraulic, or electric motor systems.

Shears are used to cut sheet metal up to $\frac{1}{4}$ in. (6.4 mm) thick. Slitting shears are used primarily to make straight cuts. Squaring shears are used to square and trim metal stock. Other types of shears are used for making zigzag cuts, slots, notches, and special design work. *Punches* are used to punch holes or small openings in the sheet metal. A *bar folder* is used to form narrow bends, folds, and hems in sheet metal. Larger bends in sheet metal are performed with brakes, known as a *cornice brake* and a *box and pan brake.*

4–1d TOOL MAINTENANCE

Good tool maintenance begins with purchasing good-quality tools that are appropriate for the work to be performed. Many low- or sale-priced tools are not made of good-quality metal and materials and will not withstand much more than a one-time use. Poor-quality is characteristic of:

- Screwdrivers that bend or break when applied to a screw
- Plier jaws that bend, break, or become loose

Chap. 4 / Basic Tools and Their Use

- Drill bits that do not drill well and break easily
- Drill chucks that do not tighten securely
- Wrenches that bend or break
- Diagonal cutters that nick easily
- Long nose pliers whose jaws break easily and other tools that are suspect by observation.

Purchasing good-quality tools is less expensive in the long run. Make sure that the size of the tool is matched to the work most frequently encountered. Staying with name brands is usually a good bet. Tools used for electronic use, such as screwdrivers, pliers, and cutters, should have insulation on the handles. If you have any doubts about which tools are best, consult an experienced electronics assembler or technician.

Basic tool maintenance procedures are given below.

1. Hacksaw
   a. Replace worn and dull blade.
   b. Discard bent or broken saw frame.
2. Pliers
   a. Repair or replace damaged handle insulation.
   b. Keep clean and lightly oil the pivot pin.
   c. Discard if it has loose or broken jaws or handles.
3. Wrenches
   a. Keep clean and rust free.
   b. Discard worn or broken tools.
   c. Keep gears on adjustable wrenches clean and lubricated.
4. Screwdrivers
   a. Regrind worn or damaged flat-blade types.
   b. Discard damaged Phillips-head type.
   c. Discard all types with damaged or broken handles.
5. Files
   a. Keep clean and free of clogged materials.
   b. Discard broken files.
6. Cutting edges
   a. Keep cutting edges on cutters, diagonal pliers, and other tools clean and sharp.
7. Power tools
   a. Repair or replace worn or frayed line cords.
   b. Check moving parts for proper motion.
   c. Check mounting parts such as chucks and blade holders for proper operation.
   d. Discard units if they have broken or cracked cases.
   e. Replace any cracked or worn bench grinder stone wheels.

4-1e GENERAL SAFETY PRECAUTIONS

Electronic assemblers and technicians are required to use hand tools and power tools to build the mechanical parts of electronic equipment. Many accidents are caused by improper and thoughtless use of tools. To protect yourself and the materials you are working on, learn to use tools correctly and pay attention to safety hazards. In this section we introduce you to some new safety rules and remind you of some old ones.

4–1e.1 Personal Safety Rules

1. When working on or near rotating machinery, secure loose clothing and tie-up long hair.
2. Make sure that power tools isolate line voltage from ground by means of an isolation transformer.
3. Make certain that the floor is electrically insulated either by asphalt tile or rubber mats and/or be sure to wear shoes with rubber soles.
4. Do not carry sharp-edged or pointed tools in your pockets.
5. Wear gloves and goggles when required.
6. Do not wear rings or jewelry when working with mechanical or electrical devices.
7. Do not defeat the purpose of any safety device, such as fuses, circuits breakers, or interlocks. Shorting across these devices could cause excessive current flow and destroy or seriously damage equipment or cause a fire.
8. Be careful with the chair you are using. Keep all legs of the chair on the floor at all times.
9. Be careful working around voltages greater than 30 volts.
10. Keep your work area clean, neat, and free of debris.
11. Do not indulge in horseplay or practical jokes in any work area.
12. Pay attention to what you are doing and have patience as you work.

4–1e.2 Hand Tool Safety Rules

1. Keep tools clean and in proper working condition.
2. Reduce hand pressure on a hacksaw before the cut is completed. The hacksaw can fall abruptly and injure the hand.
3. Do not use pliers, wrenches, and screwdrivers as a hammer.
4. Do not use long nose pliers as a wrench.
5. Take care in using long nose pliers and diagonal cutters because they can pinch and cut the hand.
6. Whenever possible, pull on a wrench; do not push it.
7. Always put a handle on a file when you use it.
8. Be especially careful when using cutting blades. Keep your other hand and body clear of cutting blades.
9. Be sure that hammer heads and screwdriver blades are fastened tightly on their handles.
10. Be careful when using a soldering iron or gun; they can burn and cause fires.
11. Place workpieces on a table or in a bench vise when using a screwdriver, wrench, pliers, or other tool. A slip of the tool can injure the hand.
12. Wear safety goggles or glasses where recommended or with the possibility of flying bits of debris.

4–1e.3 Power Tool Safety Rules

1. Make sure that the power tool is in good operating condition.
2. Make sure that line cords are not cracked or frayed.
3. Do not use tools with wet hands or feet.
4. Keep power tool safety guards in place.
5. Get instructions on tool use before operating.

6. Wear safety goggles or glasses when using power tools.
7. When drilling, fasten the workpiece securely.
8. When drilling, use a sharp drill bit and let the bit do the drilling.
9. Use sharp blades when using power saws.
10. Do not force the workpiece when using a grinder.

4–1e.4 General Workshop or Laboratory Safety Rules

1. Keep all hand tools and power tools in good working order.
2. Do not remove safety guards or devices from tools.
3. Do not turn on or operate a power tool about which you have not been instructed.
4. Turn off power tools or circuits when you have finished working on them.
5. Do not use a solvent unless you know how to use it correctly and safely.
6. Clean up spilled liquids immediately.
7. Keep the floor clean and clear of material scraps and litter.
8. Report defective tools or equipment to your supervisor or instructor.
9. Report accidents to your supervisor or instructor regardless of how small they may seem. Knowledge of small accidents can prevent large accidents.

≡≡≡ SECTION 4–2
DEFINITION EXERCISES

Write a brief description of each of the following terms.

1. Hacksaw _____

_____

2. Claw hammer _____

_____

3. Ball peen hammer _____

_____

4. Slip-joint pliers _____

_____

5. Channel-lock pliers _____

_____

6. Vise-grip pliers _____

_____

7. Bench vise _____

_____

8. C-clamp _____

_____

9. Open-end wrench _____

_____

10. Box-end wrench _____

_____

11. Adjustable wrench _____

_____

12. Crescent wrench _____

_____

13. Ratchet wrench _____

_____

14. Socket wrench _____

_____

15. Flat-blade screwdriver _____

_____

16. Phillips-head screwdriver _____

_____

17. Hex or Allen wrench _____

_____

18. Spline or Bristo wrench _____

_____

19. Utility cutter _____

_____

20. Metal files _____

_____

21. Electrician's 6-in-1 tool _____

_____

22. Wire stripper _____

_____

23. Diagonal cutters _____

_____

24. Long nose pliers _____

_____

25. Hand drill _____

_____

26. Drill bits _____

_____

27. Chuck key _____

_____

28. Hole cutter _____

_____

29. Jigsaw _____

_____

30. Drill press _____

_____

31. Grinder _____

_____

32. Shears _____

_____

33. Punches _____

_____

34. Bar folder _____

_____

35. Brakes _____

_____

Complete this section before beginning the next section.

1. Identify and label the tools shown in Figure 4-27.

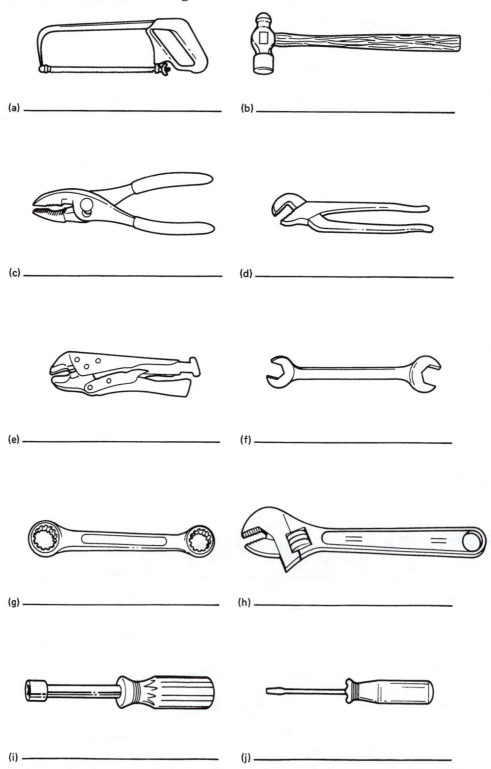

(a) _____   (b) _____

(c) _____   (d) _____

(e) _____   (f) _____

(g) _____   (h) _____

(i) _____   (j) _____

Figure 4-27

(k) _____   (l) _____

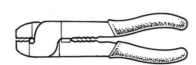

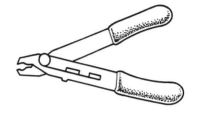

(m) _____   (n) _____

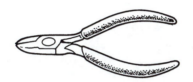

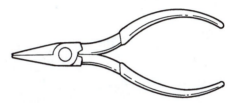

(o) _____   (p) _____

2. List the six operations or jobs that an electrician's 6-in-1 tool can perform.

   a.

   b.

   c.

   d.

   e.

   f.

3. List four steps to stripping insulation from a wire.

   a.

   b.

   c.

   d.

4. Match the operation in column A with the most correct tool to use in column B.

Column A	Column B
_____ **a.** Stripping wire	**1.** long nose pliers
	2. box-end wrench
_____ **b.** Cutting small screws	**3.** diagonal cutters
	4. nut driver
_____ **c.** Loosening "frozen" nuts	**5.** Allen wrench
_____ **d.** Cutting a $\frac{1}{2}$-inch shaft	**6.** hacksaw

_____ **e.** Bending component leads

_____ **f.** Cutting wire

_____ **g.** Driving a nut, while the screw is being held

_____ **h.** Holding a 2-inch pipe

_____ **i.** Bending a small piece of metal

_____ **j.** Tightening a setscrew

_____ **k.** Hitting a center punch

_____ **l.** Holding an object being drilled

7. 6-in-1 tool
8. channel-lock pliers
9. slip-joint pliers
10. wire stripper
11. bench vice
12. ball peen hammer

5. List the maintenance procedures for the following tools.

1. Hacksaw

   a.

   b.

2. Pliers

   a.

   b.

   c.

3. Wrenches

   a.

   b.

   c.

4. Screwdrivers

   a.

   b.

   c.

5. Files

   a.

   b.

6. Cutting edges

   a.

7. Power tools

   a.

   b.

   c.

   d.

   e.

EXPERIMENT 1. Using a Hacksaw

Caution! The small metal filings and dust caused by the use of hacksaws, drills, and files could produce "shorts" in the electrical circuits and damage equipment. This type of work should be performed away from circuit boards and test equipment such as power supplies, signal generators, oscilloscopes, and other electrical devices.

Objective:

To demonstrate the proper use of the hacksaw when cutting a shaft or pipe.

Introduction:

The proper use of any tool is important for safety reasons and to end up with a good finished workpiece. Care must be taken to get an even cut when using a hacksaw.

Materials Needed:

1 Hacksaw with sharp blade
1 Bench vise (mounted)
1 Flat metal file
1 Ruler or other measuring device
1 Pencil
1 Piece of 12-in. shaft of pipe, $\frac{1}{4}$ to $\frac{1}{2}$ in. in diameter. Soft metal such as aluminum, copper, or soft steel is recommended.

PVC pipe can also be used in this experiment to eliminate metal filings and dust.

Procedure:

1 Using the ruler and pencil, measure 1 in. from the end of the pipe and make a thin line perpendicular to the pipe's length.

2. Place the pipe into the bench vise, leaving a little more than the 1-in. mark sticking out past the end of the vise jaws. *Remember not to overtighten the vise jaws.*

3. Refer to Section 4–1a.2 and Figure 4–3 before beginning to cut the pipe. Proceed to cut the pipe on the pencil mark that you made, taking care that the cut is even and straight. The blade of a hacksaw often bends or twists, which results in a slanted or angular cut. *Take your time and be careful.*

4. After the pipe is cut, examine the cut end for metal burrs or roughness.

5. Refer to Section 4–1a.8 and Figure 4–12. Very carefully use the metal file with both hands to smooth out the metal burrs on the cut end of the pipe. *Do not overfile.*

6. Remove the pipe from the vise and examine you work.

Fill-in Questions:

1. A hacksaw is used for cutting _____

    _____ .

2. A file is used to remove metal _____

    _____ and smooth _____
    edges.

3. When sawing, the workpiece should be

    held securely in a _____ _____

    _____ .

EXPERIMENT 2. Using a Hand Drill

Objective:

To demonstrate the proper method of drilling a hole in metal.

Introduction:

Care must be taken when drilling in metal not only for safety reasons, but to have a straight and round hole as the finished job.

Materials Needed:

1 Drill bits of assorted sizes
1 Bench vise
1 Small, round metal file or at least a flat metal file
1 Center punch
1 Ball peen hammer
1 Ruler or measuring device
1 Pencil
1 Piece of $\frac{1}{8}$- to $\frac{1}{4}$-in.-thick aluminum, approximately 2 in. wide and 4 in. long.

Procedure:

1. Draw a diagonal line from one of the top corners to the opposite bottom corner. Now draw a line from the other top cor-

ner to the opposite bottom corner. The center of the metal is where the two lines cross.

2. Place the aluminum piece on a flat surface. Refer to Figure 4–21. Using the center punch and the ball peen hammer, make an indention in the aluminum piece where the two lines cross.

3. Refer to Figure 4–22. Secure the piece of aluminum in the bench vise.

4. Select a $\frac{1}{8}$- to $\frac{1}{4}$-in. drill bit and mount it in the drill chuck. Remember to tighten the chuck with the chuck key.

5. Plug the drill cord into a 120-V outlet.

6. Proceed to drill the hole in the aluminum where the indention is located. Keep the drill directly upright and steady. Be sure to check yourself as you drill. Try not to punch the hole in the aluminum with too much pressure. Go slowly and let the drill bit do the work. A small amount of light oil may be applied to the drill bit and at the hole location to facilitate the drilling process. If the drill begins to smoke, it is too dull to drill the hole properly.

7. After the hole is drilled, remove the piece of aluminum from the bench vise and look for metal burrs on the reverse side.

8. Remove any metal burrs with the file, but remember not to take off too much metal.

Fill-in Questions:

1. A starting indention is made in a workpiece to be drilled with a _____

   _____ _____ .

2. When drilling you must hold the drill upright and _____ .

3. You must not _____ the drill, but allow the bit to do the drilling.

4. Any burrs or rough metal around the hole should be removed with a _____

   _____ .

EXPERIMENT 3. Stripping Insulation from Wire

Objective:

To show the proper method for stripping insulation from wire.

Introduction:

When stripping insulation from wire it is very important not to cut any strands or nick the wire. Any strands cut or nicked will not carry the full current, thereby making the remaining wires carry more current, which could produce heat and cause a fire.

Materials Needed:

1 Wire stripper

1 Pair of diagonal cutters

1 Blade screwdriver

1 Foot-long insulated solid strand wire, size 18 to 22 gage.

2 Foot-long strips of insulated stranded wire size 18 to 22 gage.

Procedure:

1. Take a piece of the solid strand wire and look at the size of the copper wire.

2. Refer to Section 4–1b.1 and Figure 4–16a. Adjust the wire stripper so that the size of the opening is the same as the size of the copper wire.

3. Refer to Figure 4–16b. Place the stripper about $\frac{1}{2}$ in. from the end of the wire.

4. Squeeze the wire stripper handles tightly. This action cuts the insulation off the wire.

5. Pull the wire stripper off the end of the wire. The insulation should slide off.

6. Inspect the wire for any nicks, as shown in Figure 4–28. If there are any defects, you will have to adjust the wire stripper for a little-larger opening, repeating steps 2 to 6.

7. Perform the same operation on the stranded wire using steps 2 through 7.

Fill-in Questions:

1. The first operation for stripping wire is to _____ the opening of the stripper.

2. Once the stripper is placed over the wire, you must squeeze the handles tightly

to _____ the insulation.

3. To remove the cut insulation, you must

_____ the stripper off the end of the wire.

4. Any wire with nicks or missing strands

must be _____ again.

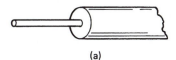
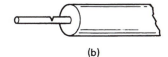

(a)   (b)

Figure 4–28 Stripped wire: (a) correct strip for solid wire; (b) incorrect strip for solid wire; (c) correct strip for stranded wire; (d) incorrect strip for stranded wire.

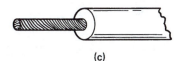
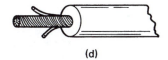

(c)   (d)

≡SECTION 4–5
INSTANT REVIEW

- The *hacksaw* is used to cut small shafts, pipes, and thin sheets of metal and plastic.
- *Hammers* are used in electronic work for driving punches, prying components apart (claw hammer), and shaping and straightening unhardened metal.
- *Slip-joint pliers* are used for gripping, turning, and bending.
- *Channel-lock pliers* are suited for gripping round, square, flat, and hexagonal objects.
- *Vise-grip pliers* have a lock feature for clamping objects.
- A *bench vise* is used for securely holding workpieces while being cut, drilled, or assembled.
- *C-clamps* are used for holding small parts together during gluing, soldering, or assembling.
- An *open-end wrench* helps to speed up tightening and loosening nuts and bolts.
- A *box-end wrench* is best to use on loosening a "frozen" nut.
- An *adjustable wrench,* sometimes called a *crescent wrench,* is suitable for working on equipment requiring different-sized nuts and bolts.
- A *ratchet wrench,* often called a *socket wrench,* is a valuable tool for speeding up work and reaching hard-to-get to places.
- A *nut driver* helps speed up work and is excellent for holding nuts while turning the screw or bolt.
- The *flat-blade screwdriver* is the standard type with a single spade type of blade.
- The *Phillips-head screwdriver* resembles a cross on the end and has better tightening qualities in some operations.
- The *hex* or *Allen wrench* is a small 90°-shaped tool used on setscrews.

- The *spline* or *Bristo wrench* is similar to the hex wrench but has more defined edges.
- The *utility cutter* has many uses for cutting, but extreme care must be exercised when using this tool.
- *Metal files* are used for removing burrs and rough edges on metal and plastic.
- The *electrician's 6-in-1 tool* can be used to:
    1. Cut wire
    2. Crimp insulated terminals
    3. Crimp noninsulated terminals
    4. Check the size of small screws and bolts
    5. Cut bolts
    6. Strip wire
- A *wire stripper* is used to remove insulation from electrical wire and can be adjusted so as not to cut or nick the wire.
- *Diagonal cutters* are used for cutting wire and component leads of electronic parts. They should not be used to cut steel wire.
- *Long nose pliers* are used for holding, bending, and as a heatsink.
- A *hand drill* is a power tool used for making holes, but must be held straight and secure while drilling.
- *Drill bits* are mounted in a drill.
- A *chuck key* is used to tighten a bit in a drill.
- A *hole cutter* has a center drill bit and a larger circular saw-type blade for making larger holes.
- A *jigsaw* is a power tool used for cutting shafts, pipes, and thin sheets of metal and plastic.
- A *drill press* is a stationary drill that provides the best situation for drilling holes requiring accuracy.
- A *bench grinder* is a power tool used to remove burrs and rough metal. It is also used to remove metal from parts that need to fit together.
- *Safety is of the utmost importance* when working with tools.
- *Tool maintenance is absolutely necessary* for safety to persons and workpieces.
- *Tools should be kept clean and* cutting edges kept *sharp.*
- *Movable parts* on tools *should be kept* lightly *oiled.*
- *Electrical cords* on power tools *that are frayed* or cracked *should be replaced.*
- *Replace any cracked* or worn *tool.*

≡≡≡ **SECTION 4–6**
SELF-CHECKING QUIZ

Circle the most correct answer for each question.

1. To cut a $\frac{1}{8}$-in. shaft, you would use a:
   a. wire stripper
   b. utility cutter
   c. hacksaw
   d. pair of diagonal cutters

2. The best tool to use for holding a nut while you turn the screw is:
   a. slip-joint pliers
   b. a nut driver
   c. a C-clamp
   d. vise-grip pliers

3. The best tool to use for loosening a "frozen" nut is:

   a. vise-grip pliers
   b. an adjustable wrench
   c. an open-end wrench
   d. a box-end wrench

4. A hex wrench could be used to remove a:

   a. control knob
   b. nut and bolt
   c. burr from metal
   d. none of the above

5. To cut a very thin sheet of cardboard you could use a:

   a. hacksaw
   b. wire stripper
   c. utility cutter
   d. pair of diagonal pliers

6. To remove burrs and rough metal from a workpiece, you would use a:

   a. utility cutter
   b. wire stripper
   c. metal file
   d. all of the above

7. To tighten a nut on a screw, you could use:

   a. an open-end wrench
   b. a box-end wrench
   c. an adjustable wrench
   d. all of the above

8. A chuck key is used with a:

   a. hand drill
   b. jigsaw
   c. ratchet wrench
   d. grinder

9. To bend the leads on a resistor you would use:

   a. a screwdriver
   b. a hex wrench
   c. vise-grip pliers
   d. a pair of long nose pliers

10. With an electrician's 6-in-1 tool you could:

    a. strip wire
    b. cut screws
    c. crimp wire terminals
    d. all of the above

11. With a worn flat-blade screwdriver, you would:

    a. just clean it
    b. discard it
    c. regrind the blade
    d. none of the above

12. With a worn Phillips-head screwdriver, you would:

    a. just clean it
    b. discard it
    c. regrind the blade
    d. none of the above

13. If you had to make an 18-in.-long cut in a $\frac{1}{4}$-in. piece of aluminum, you would use a:

    a. jigsaw
    b. hacksaw
    c. hand drill
    d. grinder

14. A good safety precaution to use with power tools is to:

    a. wear gloves
    b. hold small workpieces in your hand
    c. wear safety goggles or glasses
    d. none of the above

15. Screwdriver blades and drill bits can be sharpened with:

    a. a jigsaw
    b. a grinder
    c. diagonal cutters
    d. a utility cutter

16. To pry apart two objects, you could use:

    a. a screwdriver
    b. a file
    c. a wrench
    d. none of the above

17. You should always wear safety goggles when working with:

    a. a screwdriver
    b. a wrench
    c. power tools
    d. a nut driver

18. When using a wrench, whenever possible pull on it; do not push on it.

    a. True
    b. False

19. To save job time, any tool can be used as a hammer.

    a. True
    b. False

20. If you scratch your finger on a power tool, you do not need to report it to your supervisor.

    a. True
    b. False

ANSWERS TO FILL-IN QUESTIONS
AND SELF-CHECKING QUIZ

Experiment 1: **(1)** metal **(2)** burrs, rough **(3)** bench vise

Experiment 2: **(1)** center punch **(2)** steady **(3)** force **(4)** file

Experiment 3: **(1)** adjust **(2)** cut **(3)** slide **(4)** stripped

Self-Checking Quiz: **(1)** c **(2)** b **(3)** d **(4)** a **(5)** c **(6)** c **(7)** d **(8)** a
(9) d **(10)** d **(11)** c **(12)** b **(13)** a **(14)** c **(15)** b
(16) d **(17)** c **(18)** a **(19)** b **(20)** b

Unit 5

Hardware and Mechanical Assembly

INTRODUCTION

Hardware is usually associated with those parts used to fasten devices together, such as nuts, bolts, screws, washers, brackets, clamps, and other fasteners. *Electrical hardware* can be referred to those parts that support electrical devices such as switches, fuse holders, sockets, and terminal strips.

The quality and durability of a product is largely determined by how well its parts are mounted. Loose or carelessly mounted parts will result in an inoperative piece of equipment that may even be dangerous to use. It is important for electronic assemblers and technicians to have the proper skills in mechanical assembly.

UNIT OBJECTIVES

Upon completion of this unit, you will be able to:

1. Identify various types of screws, washers, and other hardware devices.
2. List the classifications of screws and bolts.
3. Select the proper screws, washers, and nuts for a specific job.
4. Fasten two parts together correctly using the proper hardware.
5. Explain the correct procedure for mounting a solder lug, cable clamp, grommet, stand-off, potentiometer, toggle switch, slide switch, snap-in switch, fuse holder, LEDs, and a strain relief.

FUNDAMENTAL CONCEPTS

5-1a GENERAL HARDWARE USED IN ELECTRONICS

The *chassis* is the base or framework on which electronic components are mounted using hardware. Hardware often used in electronic assembly is shown in Figure 5-1. Newer techniques in style and design have changed the way that chassis look. However, the principles are the same, and knowledge of proper assembly methods is still required.

Machine screws (Figure 5-1a, b, c, and f) are used to fasten metal and plastic parts together. *Sheet-metal* or *self-tapping screws* (Figure 5-1d and e) are used to fasten directly into a chassis or part. *Nuts* (Figure 5-1g and h) are used with machine screws and bolts for securing or fastening parts. Some nuts have a serrated washer attached to them which digs into the metal chassis or part and provides a good electrical contact. Separate *serrated washers* (Figure 5-1i and j) are placed between the nut and tightening surface and lock the nut while providing a good electrical connection. The serration may be internal (Figure 5-1i), or external (Figure 5-1j), or may be both internal and external. A *split-lock washer* (Figure 5-1k), when compressed, has a tension and locks a nut where an electrical connection is not required. The tension of the washer against the surface and nut keeps the nut from working loose when subjected to vibrations or stress. A *flat washer* (Figure 5-1l) is used as a spacer and also not to cause marring of the surface of a panel or chassis. The *fiber washer* (Figure 5-1m) is used for electrical insulating purposes. It may have a raised portion or shoulder as shown. A *solder lug* (Figure 5-1n) is used to provide a chassis or "ground" connection for wires. It may have a serrated hole or serrated washers could be used to make a good electrical connection. A *cable clamp* (Figure 5-1o) fits around a cable or group of wires for securing to a chassis. The rubber or plastic *grommet* (Figure 5-1p) is used to line the holes in metal chassis where wires pass through. This prevents any cutting of the insulation of the wires and protects against short circuits. *Solderless terminals* (Figure 5-1q and r) are

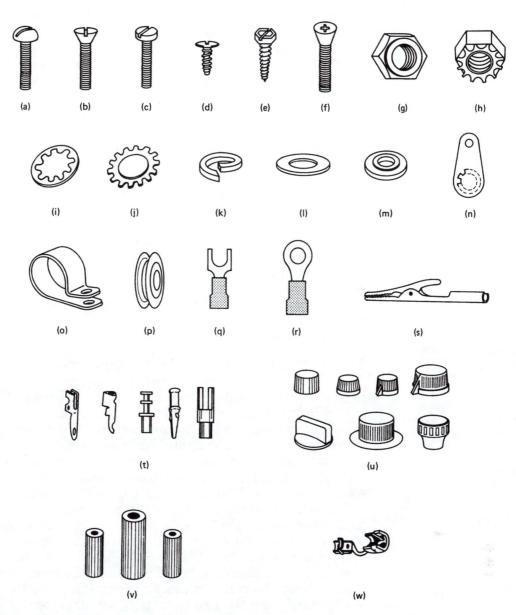

Figure 5-1 General electronic hardware: (a) round-head screw; (b) oval-head screw; (c) binder-head screw; (d) sheet-metal screw; (e) sheet-metal hex-head screw; (f) Phillips-head screw; (g) hex nut; (h) hex nut with lock washer; (i) internal-tooth lock washer; (j) external-tooth lock washer; (k) split-ring lock washer; (l) flat washer; (m) fiber washer with shoulder; (n) solder lug; (o) cable clamp; (p) rubber grommet; (q) solderless terminal—spade tongue; (r) solderless terminal-ring tongue; (s) alligator clip; (t) PC and/or vector board component mounting terminals; (u) control knobs; (v) spacers or stand-offs; (w) strain relief. (From F. Hughes, *Illustrated Guidebook to Electronic Devices and Circuits,* Prentice Hall, Englewood Cliffs, N.J., © 1981, Fig. 1–10, p. 9 Reprinted with permission.)

crimped on the ends of wires that requires screw-type terminals. An *alligator clip* (Figure 5–1s) is normally attached to each end of a wire to create a clip lead used in testing or manufacturing electronic circuits. The alligator clip may also be used as a heat sink when soldering solid-state components. *Component mounting terminals* (Figure 5–1t) are used to mount components on printed circuit boards and vector boards. They provide ruggedness and dependability and are covered in more detail in Unit 6. *Control knobs* (Figure 5–1u) of various sizes are used for switches, potentiometers, tuning capacitors, and other controls associated with circuit operation. *Spacers* or *stand-offs* (Figure 5–1v) are made of metal, phenolic, or ceramic and are used for special component mounting applications. A *strain relief* (Figure 5–1w) is usually made of plastic and fits into the hole of a chassis while applying pressure to the wire or line cord that is passing through the hole. It provides the same function as a rubber grommet while making a dependable connection for a line cord to a piece of equipment.

5–1b SCREW AND BOLT CLASSIFICATION

There are many types of fasteners used in industry, but screws and bolts remain as the conventional method for attaching two parts together. Machine screws and machine bolts are used for joining together two metal or plastic parts. The screw or bolt is placed through predrilled holes (slightly larger than the screw) in the parts, and then a nut is used to tighten the two parts together. If the parts will be subjected to vibration, shearing, or shaking, a lock washer is used under the nut. Most screws and bolts are made of steel with a plating of brass, zinc, cadmium, or chromium to resist rust and make them more attractive to the eye. Other screws and bolts are made of solid brass, bronze, aluminum, or plastic. Specific types of screw or bolts are used for various jobs.

5–1b.1 Screw and Bolt Use or Application

Figure 5–2 shows various types of screws and bolts. The *wood screw* (Figure 5–2a) cuts its own threads in the wood when turned into a counterbore (a predrilled hole a little smaller than the diameter of the screw) of the proper size. As the screw is driven down, the raised threads force the wood fibers apart, and the advance of the threads draws the parts being joined tightly together when the screw head is finally seated. Once seated, a screw will not shake loose or release easily.

A *machine screw* (Figure 5–2b) is threaded its entire length. All machine screws are slotted or have some other configuration in the head so that they may be driven by the corresponding tool.

Self-tapping screws (Figure 5–2c) have a head for driving with a screwdriver or socket wrench. A hole smaller in diameter than the screw is drilled into the metal or plastic parts. The screw threads itself into the hole and joins the two parts together when driven by the tool.

Setscrews (Figure 5–2d) are used for attaching wheels, pulleys, knobs, and other parts together, usually on a shaft. The part to be attached usually has threads tapped into it. The setscrew is placed into these threads and tightened on the shaft with either an Allen or Bristo wrench.

A *stove bolt* (Figure 5–2e), originally designed to assemble stoves, has diversified uses, even though the name remains. The head has the appropriate indentation for use with screwdrivers, and it may only be threaded a couple of inches from the end. It usually uses a square nut, although a hex nut with the same type of threads can be used.

The *machine bolt* (Figure 5–2f) has a square head and only a small number of threads on the other end. A wrench is needed to drive the machine bolt. Square or hex nuts may be used for tightening.

A *carriage bolt* (Figure 5–2g) has a smooth round head with a square shoulder beneath it. The part to be attached has a square cut hole in it. The shoulder fits into the square hole for holding while the nut is being tightened.

Stove bolts, machine bolts, and carriage bolts are used in assembling large parts which take up most of the length of the bolt before the nut is attached. Screws are normally used for flat thin pieces that are attached together.

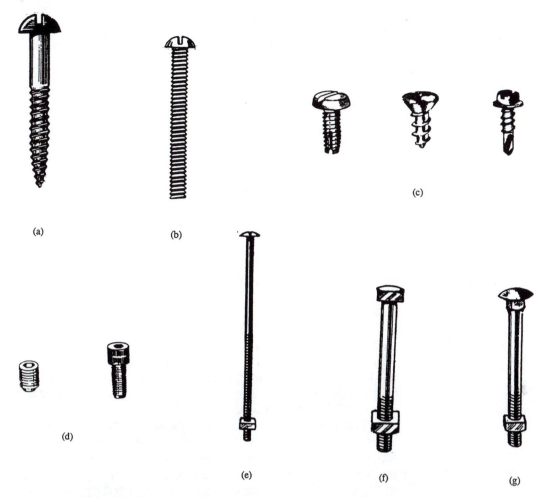

Figure 5–2 Types of screws and bolts: (a) wood screw; (b) machine screw; (c) self-tapping screws used on metal and plastic; (d) setscrews; (e) stove bolt; (f) machine bolt; (g) carriage bolt.

5–1b.2 Type of Head

Screws and bolts have different types of heads, which are used for specific holding purposes. It is important to use the proper head for the specific job to have a good-quality product and also to ensure safety to personnel and equipment.

Figure 5–3 shows several types of screw heads. The *round-head screw* (Figure 5–3a) is probably the most common type and provides easy driving and maximum strength at the head. The *truss-head screw* (Figure 5–3b) has a smaller head in height but a little larger holding area. A *fillister-head screw* (Figure 5–3c) has edges that are more abrupt rather than sharp, and are excellent for mounting front panels. The *binding-head screw* (Figure 5–3d), similar to the fillister-head screw, has a smaller head and is used for mounting brackets and internal parts. An *oval-head screw* (Figure 5–3e) uses a countersunk hole. A *countersunk hole* is a regular drilled hole which is then drilled with a larger bit to provide a funnel-shaped hole leading to the regular-size hole opening. A *washer-head screw* (Figure 5–3f) has an extended area on it similar to a washer. This screw head provides maximum holding strength over a larger area and is often used on front panels. The *pan-head screw* (Figure 5–3g) has a fairly large holding area and is often found on self-tapping screws. The *flat-head screw* (Figure 5–3h) is used on parts where the head must be flush with the surface and requires a countersunk hole.

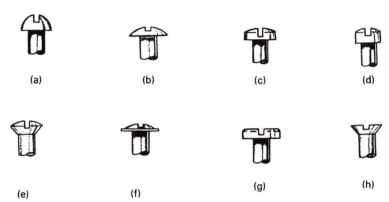

Figure 5–3 Types of screw heads: (a) round-head; (b) truss-head; (c) fillister-head; (d) binding-head; (e) oval-head; (f) washer-head; (g) pan-head; (h) flat-head.

5–1b.3 Gage and Size

The diameter and length of a screw or bolt is important when assembling parts. Using the correct-size screw prevents loose parts and maximum holding surface area for rigidly held parts. Figure 5–4a shows the American Screw Standard Gage for screws and bolts.

The gage number corresponds to the diameter of the screw. The larger the gage number, the larger the diameter. For example, a No. 6 screw is about $\frac{9}{64}$ in. in diameter, while a No. 14 screw is about $\frac{15}{64}$ in. in diameter. Screws are also classified by the type of threads, as shown in Figure 5–4b. The type of thread is indicated by the number of threads in 1 in. of the screw. For example, the often-used 6–32 screw has 32 threads per inch.

There are several standard types of screws relating to the gage, number of threads per inch, the angle of the cut of the threads, whether the edges of the threads are sharp or rounded, and various other design considerations. Normally, electronic assembly work involves three main types of screw standards: the American National Course Thread (NC), the American National Fine Thread (NF), and the International Standards Organization (ISO), which uses metric dimensions in millimeters. The course thread and the fine thread screws have the same diameter size, but the fine thread standard has more threads per inch. Table 5–1 shows a comparison of NC and NF threads and indicates the nearest diameter for metric sizes.

TABLE 5–1
Thread Size Comparisons

Diameter (in.)	NC Gage and Threads/inch	NF Gage and Threads/inch	Metric (mm)
0.073	1 × 64	1 × 72	1.8542
0.086	2 × 56	2 × 64	2.1844
0.099	3 × 48	3 × 56	2.5146
0.112	4 × 40	4 × 48	2.8448
0.125	5 × 40	5 × 44	3.1750
0.138	6 × 32	6 × 40	3.5052
0.164	8 × 32	8 × 36	4.1656
0.190	10 × 24	10 × 32	4.8260
0.216	12 × 24	12 × 28	5.9864

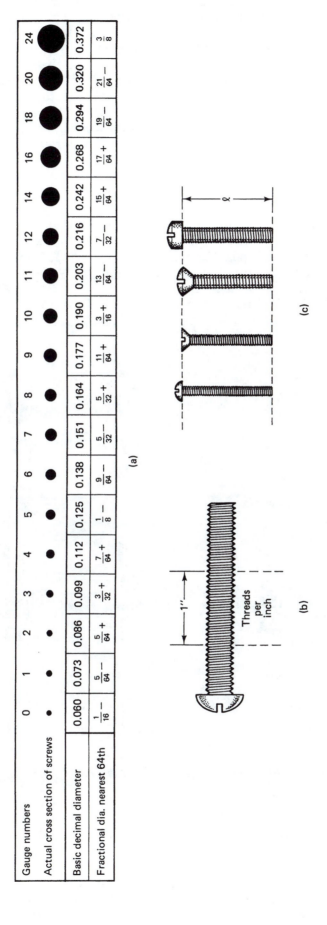

Gauge numbers	0	1	2	3	4	5	6	7	8	9	10	11	12	14	16	18	20	24
Actual cross section of screws																		
Basic decimal diameter	0.060	0.073	0.086	0.099	0.112	0.125	0.138	0.151	0.164	0.177	0.190	0.203	0.216	0.242	0.268	0.294	0.320	0.372
Fractional dia. nearest 64th	$\frac{1}{16}$ −	$\frac{5}{64}$ −	$\frac{5}{64}$ +	$\frac{3}{32}$ +	$\frac{7}{64}$ +	$\frac{1}{8}$ −	$\frac{9}{64}$ −	$\frac{5}{32}$ −	$\frac{5}{32}$ +	$\frac{11}{64}$ −	$\frac{3}{16}$ +	$\frac{13}{64}$ −	$\frac{7}{32}$ −	$\frac{15}{64}$ +	$\frac{17}{64}$ +	$\frac{19}{64}$ −	$\frac{21}{64}$ −	$\frac{3}{8}$

(a)

(b)

(c)

Figure 5-4 Screw and bolt size gage: (a) chart showing relative diameter of screws; (b) indication of threads per inch; (c) method of determining length for various screw heads.

The overall length of a screw is not the same for types of heads, as shown in Figure 5–4c. On round-head and binder-head screws the length begins directly beneath the head and runs to the end. The flat-head screw's length is its total length and an oval-head screw's length begins at the sloping or bevel point and runs to the end. It is important to acquire the proper-length screw or bolt for assembling parts. Obviously, if the screw is too short, it will not be able to fasten the parts, but if the screw is too long, it could interfere with the assembly of other parts and become dangerous to personnel working on the equipment. *A screw with proper length should be long enough to extend only one or two threads past the end of the nut.*

5–1b.4   Types of Screw Drives

The type of drive used on the head of a screw is determined by the use and application of the screw. The most common types of drives are the blade, Phillips, Allen, Bristo, and clutch types used on automotive vehicles. Figure 5–5 shows some of the types of screw drives and their variations in use today. The electronics assembler and technician should be familiar with the various types of drives that exist.

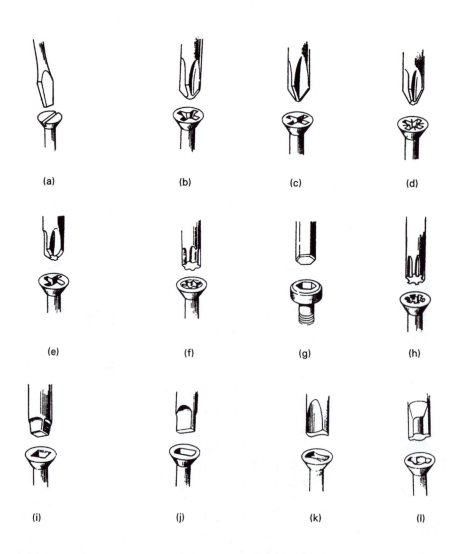

Figure 5–5   Types of screw drives: (a) flat-blade; (b) Phillips; (c) Reed & Prince; (d) Pozidriv; (e) torque set; (f) Torx; (g) hex or Allen; (h) spline or Bristo; (i) Scrulox; (j) slab; (k) clutch head type A; (l) clutch head type G.

5–1b.5 Types of Nuts

Various types of nuts are used in the assembling of parts, depending on their strength, holding capability, economical use, and accessibility. Figure 5–6 shows various types of nuts.

The standard *square nut* (Figure 5–6a) and *hex nut* (Figure 5–6b) are the most used types in assembling parts. The *interference* or *stop nut* (Figure 5–6c) is slightly smaller on one end, and as the screw is driven in, it becomes tighter and forms a locking action. The stop nut does not require a locking washer. The *cap nut* (Figure 5–6d) is used for external parts such as panels and provides a good appearance.

The *chassis mount nut* (Figure 5–5e) has a small inner shoulder that is serrated on its edges. This serrated part of the nut is pressed into an exact-size hole in the material, such as a cover or plate. The nut holds to the material and a screw may be inserted and tightened. This type of nut speeds up assembly time and eliminates the extra loose parts from being lost.

The *spring* or *speed nut* (Figure 5–6f) is made from spring steel with a special cut hole and a manufactured bend placed in it. This is an inexpensive way to produce parts and goods that most industries try and utilize. The speed nut may just be held in place by the screw passing through a hole. The nut may fit into the metal with a clipping action and remain on the metal so that the screw can be removed without loosing the nut.

Knurled thumb nuts (Figure 5–6g) are used on screws where assembly and disassembly of parts is frequent. The serrated part of the nut makes it easy to tighten and loosen with the fingers. The *wing nut* (Figure 5–6h) is used for the same reasons as the knurled thumb nut, except that more torque can be applied with the fingers for better holding action.

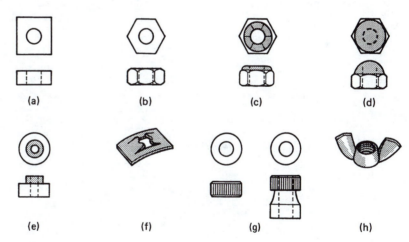

Figure 5–6 Types of nuts: (a) square; (b) hexagonal (hex); (c) interference or stop nut; (d) cap nut; (e) chassis mount nut; (f) spring or speed nut; (g) knurled thumb nuts; (h) wing (thumb) nut.

5–1c STANDARD MOUNTING PROCEDURES

To ensure a product's durability the method by which it is assembled is very important. There are certain assembly practices that apply to the mounting of regular hardware and electrical hardware. The use of the correct washer for a specific job is critical in many instances. Split-ring lock washers should be used beneath the nut when joining any two metal parts. Serrated washers should always be used under a flat washer or nut to form a good electrical connection. Flat washers are used on soft materials where you do not want to mar the surface or tear the part. The following mounting procedures of the various components will familiarize you with proper assembly methods.

5–1c.1 Mounting Two Flat Pieces

The standard mounting procedure for mounting two metal parts is shown in Figure 5–7a. The hole drilled in the parts must be a little larger than the diameter of the screw, so the screw can fit easily through both parts. The screw is first pushed through the part to be attached and then through the stationary part. A split-ring lock washer is placed on the end of the screw and then the nut is tightened on the end of the screw. The nut can be held with a wrench or pair of pliers while the screw head is turned with a screwdriver, or the screw head can be held with a screwdriver and the nut tightened with a wrench. The size of the screw should be of proper length so that only one to two threads extend beyond the end of the nut. The nut should be tight—but not overly tightened so as to break off the screw. Care and discretion must be used when working with tools.

5–1c.2 Mounting a Solder Lug

An example of a good required electrical connection is the mounting of a solder lug as shown in Figure 5–7b. The screw is pushed through the

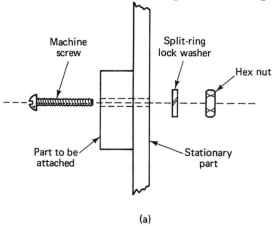

(a)

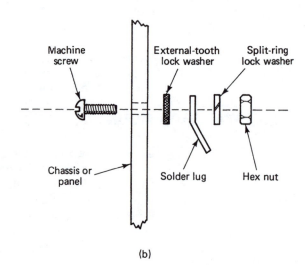

Figure 5–7 Standard parts mounting; (a) mounting one flat part to another; (b) mounting a solder lug requiring good electrical connection.

(b)

chassis or panel. A serrated external-tooth lock washer, solder lug, and split-ring lock washer is placed on the screw. The nut is then applied and tightened. As the nut tightens the components together, the serrated lock washer digs into the metal of the chassis and the solder lug, making a good electrical connection. While the nut is being tightened, the solder lug may have to be positioned in a direction that does not interfere with other components and/or accommodates neat wiring practices.

5-1c.3 Mounting a Cable Clamp

Cable clamps are used to secure wires and keep them in an orderly manner inside the chassis. Figure 5–8 shows how a cable clamp is attached to a chassis or panel. The screw is pushed through the hole in the chassis, and the cable clamp, flat washer, and split-ring washer are placed over the end. Most cable clamps are made of plastic and the flat washer protects the soft plastic from tearing against the split-ring washer when the nut is tightened. Some cable clamps have a metal surface on the outer side of the hole and a flat washer is not needed.

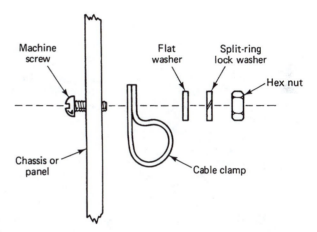

Figure 5–8 Mounting a cable clamp.

5-1c.4 Mounting a Grommet

Grommets provide insulated linings around holes in a chassis or panel where wires pass through. The hole for a rubber grommet, shown in Figure 5–9a, must be no larger than the diameter of the inside slot of the grommet. This measurement is indicated by the letter d. The hole is drilled and all burrs or rough edges are removed. The rubber grommet is then pressed through the hole until one side of the rubber edge is all the way through and the grommet fits evenly in the hole. Figure 5–9b shows how a plastic or mylar flexible strip is used as lining.

5-1c.5 Mounting Spacers and Stand-offs

Very often, parts, subchassis, and panels must be mounted above the main chassis or panel. This condition is accomplished with spacers or stand-offs as shown in Figure 5–10. A typical spacer is not threaded and must use a screw, washer, and nut for assembly. A typical stand-off is internally threaded at each end or its entire length. The stand-off is attached to the main or lower chassis with a screw, flat washer, and split-ring lock washer. The stand-off is held with a pair of pliers or wrench while the screw is turned. The upper chassis is assembled first with a flat washer, then the chassis, a split-ring lock washer, and finally, the screw. Again, the stand-off is held with a pair of pliers or wrench while the screw is turned.

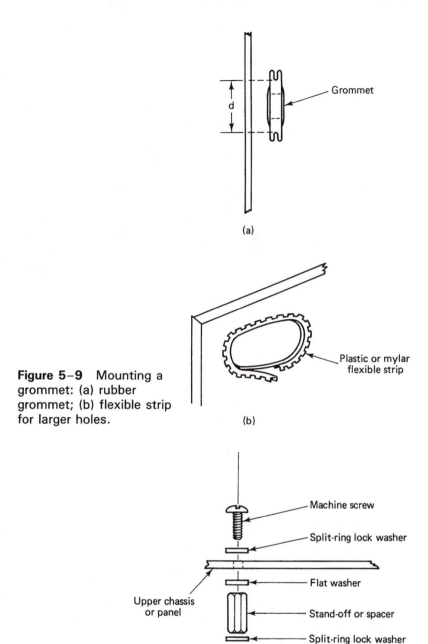

(a)

Figure 5-9 Mounting a grommet: (a) rubber grommet; (b) flexible strip for larger holes.

Grommet

Plastic or mylar flexible strip

(b)

Machine screw

Split-ring lock washer

Flat washer

Upper chassis or panel

Stand-off or spacer

Split-ring lock washer

Flat washer

Lower chassis or panel

Machine screw

Figure 5-10 Mounting a stand-off (or spacer).

5-1c.6 Mounting a Potentiometer and Knob

Most electronic equipment, such as radios, television receivers, and other products, use potentiometers in the form of volume, brightness, contrast, color, and other controls. Figure 5–11 shows a typical mounting procedure for a potentiometer.

First, a serrated lock washer is placed on the threaded shaft of the potentiometer, and then it is placed through the hole in the chassis or panel. On the other side of the panel a flat washer is placed on the protruding threaded shaft of the potentiometer. The hex nut is then tightened on the threaded shaft. The potentiometer is held in the desired position

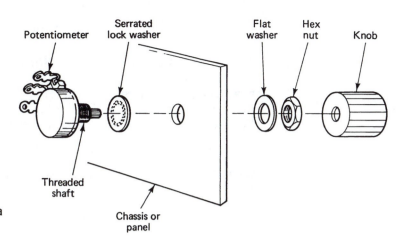

Figure 5–11 Mounting a potentiometer.

with one hand as the hex nut is tightened with a wrench or nut driver. The control knob is then slid over the movable shaft of the potentiometer and is held in place by a snug fit or with a setscrew. If the control knob has a line or some other indicator, the movable shaft is rotated counterclockwise until it stops; then the knob is placed on the shaft with the indicator at the lower left position, around the 7 o'clock point on a watch.

5–1c.7   Mounting a Toggle Switch

A toggle switch, shown in Figure 5–12, is mounted similar to a potentiometer, except that it may have a back hex nut. This back hex nut is adjusted so that there is only one thread showing past the front hex nut after the assembly is tightened. This makes a pleasing front appearance.

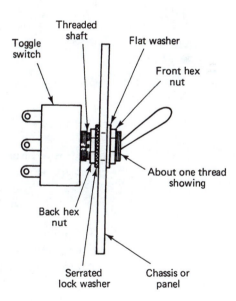

Figure 5–12   Mounting a toggle switch.

The assembly procedure is thus as follows: The back hex nut is placed on the threaded shaft of the switch, a serrated lock washer is placed on the threaded shaft, the switch assembly is pushed through the hole in the panel, a flat washer is placed on the threaded shaft, and finally, a hex nut is tightened on the threaded shaft. A thin knurled thumb nut may be used instead of the hex nut, but normally it needs to be tightened more than with just hand strength—perhaps with a pair of pliers. Be careful that the pliers do not slip and strip the ridges on the thumb nut.

5–1c.8 Mounting a Slide Switch

A slide switch needs a special rectangular hole cut in the chassis or panel to accommodate the moving mechanism, and screw holes for mounting the switch are drilled above and below this hole. Figure 5–13 shows how to mount a slide switch. The slide switch movable mechanism is pushed through the rectangular hole and then a screw is pushed through the panel on the other side so that a split-ring lock washer and hex nut can be used to attach the switch to the panel. The other screw is then attached to the panel and switch. Before the final tightening of the screws, adjust the switch so that its movable mechanism will move freely in the rectangular hole.

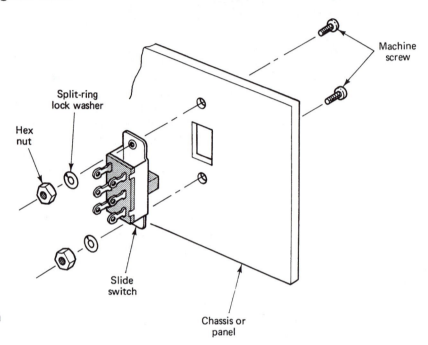

Figure 5–13 Mounting a slide switch.

5–1c.9 Mounting a Snap-in Switch

In the case of a snap-in switch, shown in Figure 5–14, a rectangular hole must be cut in the panel just large enough to allow the back side of the switch to slide through. The switch is simply pushed into the hole until the spring clips pop up behind the rear of the panel to hold it in place.

5–1c.10 Mounting a Fuse Holder

To mount a fuse holder as shown in Figure 5–15, a round hole must be cut in the panel just large enough to allow the plastic threads to pass. An external-tooth serrated lock washer is placed over the threads, and then a hex nut is used to tighten the fuse holder to the panel. A fuse is placed in the holder from the front end and then the fuse cap is tightened on the holder.

5–1c.11 Mounting LEDs

LEDs are used quite extensively in electronic products. They have replaced more bulky pilot lamps as off/on indicators and are used in many other applications. Figure 5–16 shows various methods used to mount LEDs. An LED may simply have its leads pushed through holes in a PC board (Figure 5–16a) and soldered in place to the circuit on the board. One large hole in a PC board will allow the leads to slide all the way

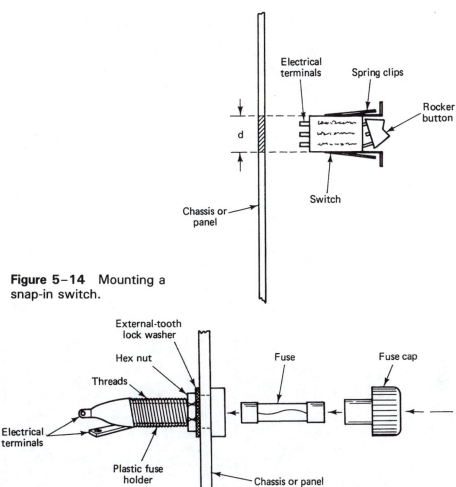

Figure 5–14 Mounting a snap-in switch.

Figure 5–15 Mounting a fuse holder.

through for a flat-mounted LED (Figure 5–16b). The leads of the LED may be bent 90° and inserted into two holes in a PC board (Figure 5–16c). Very often a panel-mounted LED will be placed into a plastic or rubber grommet already in place on the panel (Figure 5–16d). In other cases, epoxy is used to hold the LED in place on a panel (Figure 5–16e). Sometimes LEDs are mounted to a PC board that is very close to the front panel. The LEDs are then able to be pushed through holes in the panel (Figure 5–16f).

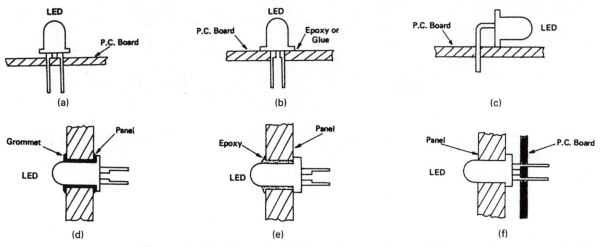

Figure 5–16 Mounting LEDs: (a) two holes in PC board; (b) one hole in PC board; (c) bent leads; (d) through panel with a grommet; (e) through panel with epoxy; (f) through panel with LED connected to PC board.

5-1c.12 Mounting a Strain Relief

A strain relief is used to reduce the tension and keep from breaking the line cord, which passes through a chassis or panel. The strain relief is usually made out of plastic. Figure 5–17 shows how to mount a strain relief.

The hole drilled in the chassis must be smaller in diameter than the body part of the strain relief which passes though the hole (Figure 5–17a). The strain relief has a removable top that can be depressed to allow it to pass through the hole. The line cord is passed through the hole first. The top of the strain relief is opened and the line cord on the outside of the chassis is placed into it (Figure 5–17b). The top is placed back into position on the strain relief and the entire assembly is pushed into the hole. A pair of long nose pliers may be needed to compress the strain relief as it goes into the hole (Figure 5–17c). The finished assembly of the strain relief will appear as shown in Figure 5–17d.

Figure 5–17 Mounting a strain relief: (a) dimensions of hole; (b) placing wire; (c) inserting strain relief; (d) finished strain relief.

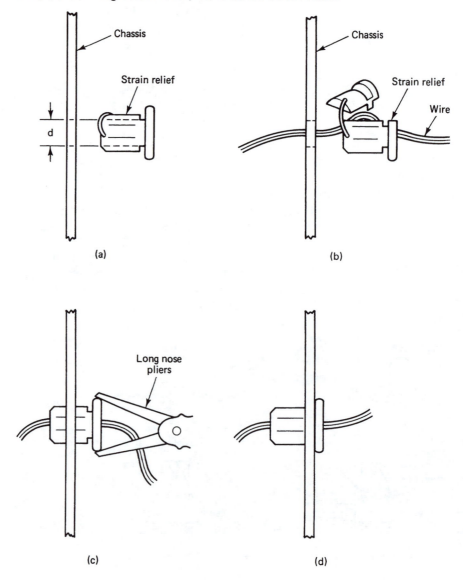

Write a brief description of each of the following terms.

1. Hardware _____

_____

2. Electrical hardware _____

_____

3. Chassis _____

_____

4. Machine screw _____

_____

5. Sheet-metal screw _____

_____

6. Self-tapping screw _____

_____

7. Setscrew _____

_____

8. Stove bolt _____

_____

9. Machine bolt _____

_____

10. Carriage bolt _____

_____

11. Square nut _____

_____

12. Hex nut _____

_____

13. Interference or stop nut _____

_____

14. Cap nut _____

_____

15. Chassis mount nut _____

_____

16. Spring or speed nut _____

_____

17. Knurled thumb nut _____

_____

18. Wing nut _____

_____

19. Serrated washer _____

_____

20. Split-lock washer _____

_____

21. Flat washer _____

_____

22. Fiber washer _____

_____

23. Solder lug _____

_____

24. Cable clamp _____

_____

25. Grommet _____

_____

26. Solderless terminals _____

_____

27. Alligator clip _____

_____

28. Component mounting terminal _____

_____

29. Control knob _____

_____

30. Spacer _____

_____

31. Stand-off _____

_____

32. Strain relief _____

_____

33. Countersunk hole _____

_____

≡≡≡ SECTION 5–3
EXERCISES AND PROBLEMS

Complete this section before beginning the next section.

1. Refer to Figure 5–2 and draw the physical appearance of the following.

    a. machine screw   **b.** self-tapping screw

    c. setscrew   **d.** stove bolt

2. Sketch the following types of screw heads (refer to Figure 5–3).

    a. round-head   **b.** truss-head   **c.** Fillister-head

    d. binding-head   **e.** oval-head   **f.** washer-head

    g. pan-head   **h.** flat-head

3. List the approximate diameter and number of threads per inch for the following screws.

	Screw No.	Diameter	Threads/inch
a.	6–32		
b.	8–32		
c.	10–24		
d.	4–40		
e.	2–56		

4. Match the application in column A with the proper nut in column B.

Column A

_____ **a.** A nut used on a panel to give a good finish

_____ **b.** A nut removed often, but must have good torque

_____ **c.** An inexpensive nut made of spring steel

_____ **d.** Standard machine screw nut

_____ **e.** A nut tightened with the fingers

_____ **f.** A nut that does not require a locking washer

_____ **g.** A nut mounted in a chassis

Column B

1. hex nut
2. stop nut
3. cap nut
4. speed nut
5. wing nut
6. knurled thumb nut
7. chassis mount nut

5. List the proper order for assembling the hardware for the following parts. Refer to the corresponding figure, also given. Part (a) has been completed as an example.

a. Two plates—Fig. 5–7a

screw _____

split-ring lock washer _____

hex nut _____

b. Solder lug—Fig. 5–7b

_____

_____

_____

_____

_____

c. Cable clamp—Fig. 5–8

_____

_____

_____

_____

d. Grommet—Fig. 5–9

_____

_____

e. Stand-off—Fig. 5–10

_____

_____

_____

_____

_____

_____

f. Potentiometer—Fig. 5–11

_____

_____

_____

_____

_____

g. Toggle switch—Fig. 5–12

_____

_____

_____

_____

_____

h. Slide switch—Fig. 5–13

_____

_____

_____

_____

_____

i. Fuse holder—Fig. 5–15

_____

_____

_____

≡≡≡ **SECTION 5–4**
EXPERIMENTS

EXPERIMENT 1. Standard Parts Mounting

Objective:

To develop basic skills in layout, drilling, and assembling two metal parts.

Introduction:

In this experiment you will have to lay out two pieces of metal to be drilled with holes and assemble the required hardware. The types of screws used can be round-head, fillister-head, or pan-head.

Materials Needed:

2 Pieces of metal (_Suggestion:_ one piece of $\frac{1}{2}$-in.-thick aluminum 1 in. wide by 3 in. long and one piece of $\frac{1}{8}$-in.-thick aluminum 3 in. wide by 5 in. long)

2 6-32 metal screw 1 in. long

2 6-32 hex nut

2 6-32 split-ring lock washer

2 6-32 external or internal tooth lock washer

1 Solder lug

1 Hand drill or drill press

1 $\frac{3}{16}$-in.-diameter drill bit

1 Center punch

1 Ball peen hammer

1 Ruler or scale measuring device

1 Pencil

1 Flat metal file

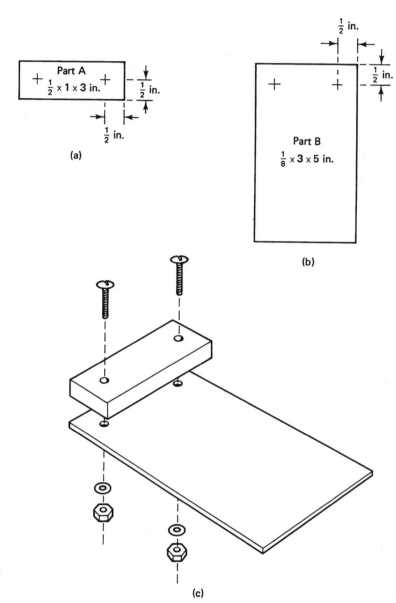

Figure 5–18 Standard parts assembly: (a) part A; (b) part B; (c) assembly drawing.

Procedure:

1. Lay out and mark with a pencil the two metal pieces according to Figure 5–18a and b.
2. Lightly center punch the cross marks on the metal pieces with the ball peen hammer.
3. Assemble the drill and bit and drill all four holes carefully. Make sure that no metal burrs are left on the metal by using the file.
4. Assemble the screws, washers, and hex nuts as shown in Figures 5–18c and 5–7a.
5. Inspect your work.
6. Disassemble one of the screws.
7. Assemble the solder lug and associated hardware as shown in Figure 5–7b.
8. Inspect your work.

Fill-in Questions:

1. To prevent a nut from becoming loose on a screw, you would use a _____ washer.

2. When a good electrical connection is required in assembling parts, you would use a _____ lock washer.

EXPERIMENT 2.   Mounting a Potentiometer

Objective:

To demonstrate the proper method for mounting a potentiometer in a panel.

Introduction:

Through panel mounting of components requires the correct-size hole for the particular component to be mounted. There are various-size potentiometers; therefore, the hole to be drilled in the panel must be a little larger in diameter than the size of the threads that go through the hole. Refer to Figure 5–11.

Materials Needed:

1 $\frac{1}{16}$- to $\frac{1}{8}$-in.-thick piece of metal approximately 3 by 5 in. or larger

1 Potentiometer

1 Serrated lock washer

1 Flat washer

1 Hex nut

1 Knob

1 Hand drill or drill press

1 Assorted drill bits

1 Center punch

1 Ball peen hammer

1 Flat metal file

1 Nut driver (same size as hex nut)

1 Hex or Allen wrench (same size as setscrew in knob if one is used)

Procedure:

1. Find the proper-size drill bit that is a little larger in diameter than the threads on the potentiometer.

2. Lay out the mark on the metal where the potentiometer is to be located.

3. Using the center punch and ball peen hammer, make a small indentation on the mark.

4. Assemble the drill and bit and drill the hole in the metal. Use the file to remove any metal burrs from around the hole.

5. Place the potentiometer through the hole in the metal as shown in Figure 5–11 and assemble the hardware to hold it in place. Hold the potentiometer with the electrical terminals in the upright position as you tighten the hex nut with the nut driver. If the knob is of the push-on type, place it on the turning shaft. If a setscrew is used with the knob, you may have to loosen it some so that it will fit on the shaft.

6. Inspect your work.

Fill-in Questions:

1. A _____ lock washer is usually placed between the potentiometer and panel when it is assembled.

2. A flat washer is placed between the panel and hex nut when mounting a potentiometer so as not to _____ the panel.

≡≡≡ **SECTION 5–5**
INSTANT REVIEW

- *Hardware* consists of nuts, bolts, screws, brackets, clamps, washers, and other types of fasteners.
- *Electrical hardware* consists of switches, fuse holders, sockets, terminal strips, and other component mounting devices.
- A *chassis* is the base or framework on which electronic components are mounted using hardware.
- *Machine screws* are used to fasten metal or plastic parts together.
- *Sheet-metal* or *self-tapping screws* are used to fasten directly into a chassis or part.
- *Setscrews* are used for attaching wheels, pulleys knobs, and other parts on a shaft.
- A *stove bolt* has a slotted head and is turned with a screwdriver.
- A *machine head* has a square head and should be turned with a wrench.
- A *carriage bolt* has a smooth round head with a square shoulder beneath it. The shoulder fits into a square hole and the bolt is tightened with a wrench on the nut end. Screws are classified by their

type of driving head, such as round-head, truss-head, fillister-head, binding-head, oval-head, washer-head, pan-head, and flat-head.

- The size of a screw is determined by its diameter or *gage.* The larger the gage number, the larger the diameter.
- Screws are also classified by the number of *threads per inch.* For example, a 6–32 screw is about $\frac{9}{64}$ in. in diameter and has 32 threads per inch.
- There are many types of screws and drivers, the most common being *flat-head, Phillips-head,* and *Allen-head.*
- A *square nut* or *hex nut* is standard for most screws and bolts.
- A *stop nut* has tapered threads and the screw will reach a point where it will stop as it is tightened. This type of nut does not require a lock washer.
- A *cap nut* provides a finished look to a piece of equipment.
- A *chassis mount nut* is pushed into a hole in the metal chassis or panel, where it remains.
- A *spring* or *speed nut* is an economical nut made from spring steel and is used in many products.
- *Knurled thumb nuts* allow fast assembly and disassembly.
- *Wing nuts* have the same function as knurled thumb nuts, but more torque can be applied to them.
- *Serrated lock washers* are used to provide a good electrical connection.
- A *split-ring lock washer* is used to make a secure mechanical connection where an electrical connection is not required.
- A *flat washer* is used to prevent scratching or damaging the surface of a panel.
- A *fiber washer* is used where electrical insulation is required between two parts.
- The *solder lug* is used for making electrical connections and is mounted to a chassis or panel.
- A *cable clamp* is used to hold or secure wires and cables.
- A *grommet* is placed in the hole of a metal chassis to prevent the cutting of the insulation of wires that are passed through the hole.
- *Solderless terminals* are crimped on wires, which are then attached to chassis terminals with screws.
- *Alligator clips* are placed on wires for making clip leads, which can be used to connect various parts of a circuit temporarily.
- *Component mounting terminals* are usually found on printed circuit boards to which electronic components are soldered.
- *Control knobs* are placed on the turning shafts of potentiometers, switches, and other devices.
- A *spacer* is used to separate two compounds with a screw, washers, and a nut.
- A *stand-off* has the same function as a spacer, but usually has internal threads at each end.
- A *strain relief* is used to protect a line cord that passes through a chassis or panel.
- A *countersunk hole* is a regular-size drilled hole with its outer edges beveled to accommodate a flat-head screw.

Circle the most correct answer for each question.

1. The type of fastener to use to connect two thin pieces of metal together is a:

   a. machine screw   **b.** machine bolt

   c. setscrew   **d.** carriage bolt

2. A screw with a diameter of about $\frac{3}{16}$ in. and 24 threads per inch would be classified as a:

   a. 24-6 screw   **b.** 10-24 screw

   c. 8-24 screw   **d.** 4-24 screw

3. If the head of the screw must be flush with the surface of a panel, you would use a:

   a. truss-head screw   **b.** pan-head screw

   c. flat-head screw   **d.** washer-head screw

4. To attach a knob on the shaft of a potentiometer you would use a:

   a. flat-head screw   **b.** truss-head screw

   c. binding-head screw **d.** setscrew

5. A sheet-metal screw can be used to fasten:

   a. a fuse holder to a panel

   b. an LED to a PC board

   c. two thin sheets of metal together

   d. a strain relief in a chassis

6. To mount a component above the main chassis you could use a:

   a. cable clamp   **b.** stand-off

   c. wing nut   **d.** carriage bolt

7. To secure wires and keep them in place on a chassis you would use a:

   a. cable clamp   **b.** spacer

   c. alligator clip   **d.** strain relief

8. To mount a toggle switch in a panel you would use the following procedure:

   a. push the switch through the hole, then place a serrated lock washer on the threads and the hex nut.

   b. push the switch through the hole, then place a flat washer on the threads and the hex nut.

   c. place a serrated lock washer on the threads, push the switch through the hole, then place a flat washer on the threads and the hex nut.

   d. place a flat washer on the threads, push the switch through the hole, then place a serrated lock washer on the threads and the hex nut.

9. Mounting a typical slide switch requires at least this many holes in the panel:

   a. 1   **b.** 2

   c. 3   **d.** 4

10. The number of screws you would use to mount a snap-in switch is:

    a. 3   **b.** 2

    c. 1   **d.** 0

11. To prevent a screw head from scratching the surface of a panel you would use a:

    a. split-ring lock washer   **b.** grommet

    c. flat washer   **d.** none of the above

12. A nut that does not require a lock washer is the:

    a. square nut   **b.** stop nut

    c. hex nut   **d.** none of the above

13. The nut that gives the best appearance to a finished panel is the:

    a. wing nut   **b.** hex nut

    c. stop nut   **d.** cap nut

14. The most economical nut used to assemble products is the:

    a. stop nut   **b.** hex nut

    c. square nut   **d.** none of the above

15. A strain relief is used to:

    a. relieve pressure on electronic components

    b. mount a fuse holder

    c. pass a line cord through a chassis

    d. all of the above

16. To protect a group of wires going through a hole in the chassis you would use a:

    a. cable clamp   **b.** grommet

    c. strain relief   **d.** all of the above

17. The back hex nut mounted on a toggle switch is used:

    a. for added strength to the assembly

    b. to adjust the length of the threads so that only one thread is visible in the front after the assembly is tightened

    c. to pull the serrated lock washer tight against the back of the panel

    d. all of the above

18. A countersunk hole is used for the:

    a. binding-head screw   **b.** LED

    c. flat-head screw   **d.** stand-off

19. The good way to mount an LED in a PC board is:

    a. flat on the PC board

    b. a slight distance above the PC board

    c. with a 90° bend in its leads

    d. all of the above.

20. A LED must always use a grommet when it is mounted in a panel.

    a. True   **b.** False

ANSWERS TO FILL-IN QUESTIONS AND SELF-CHECKING QUIZ

Experiment 1:   **(1)** lock   **(2)** serrated

Experiment 2:   **(1)** serrated   **(2)** scratch or damage

Self-Checking Quiz: **(1)** a   **(2)** b   **(3)** c   **(4)** d   **(5)** c   **(6)** b   **(7)** a   **(8)** c
    (9) c   **(10)** d   **(11)** c   **(12)** b   **(13)** d   **(14)** d   **(15)** c
    (16) b   **(17)** b   **(18)** c   **(19)** d   **(20)** b

Unit 6

SOLDERING TECHNIQUES

INTRODUCTION

Soldering is the process of joining two metals together in a solid connection or bond with a third metal alloy, known as *solder*. The technique of soldering was first used by the ancient Egyptians in the making of jewelry. Today, soldering is used extensively in such electronic devices and product areas as automobile traffic control, aviation, police, medical emergency, radio and television, computers, home entertainment, and test equipment. Because the safety of human life so often depends on electronics, it is important that these devices be highly reliable; therefore, solder connections, which are a principal component in the production of electronic equipment, must also be highly reliable.

UNIT OBJECTIVES

Upon completion of this unit, you will be able to:

1. Describe the composition of solder.
2. Explain the reason for using solder flux.
3. Identify basic soldering tools.
4. Prepare a soldering iron properly for soldering.
5. Use proper soldering methods.
6. Recognize acceptable and unacceptable solder joints.
7. Remove solder correctly from a joint.
8. Solder various types of terminals.
9. Explain the reason for using various soldering iron tips and temperatures.
10. Describe the use of a soldering gun.

6-1a   SOLDER

6-1a.1   Reasons for Soldering

Electrical circuits can be joined by mechanical means with nuts and bolts, rivets, staples, or some other type of mechanical fasteners. However, these joints can become loose from vibration and other mechanical shock. Also, when metals are exposed to air an *oxide* forms on their surface, similar to rust on iron. The oxide acts as an electrical insulator, interrupting circuit operation. Soldering overcomes these drawbacks and provides two important results:

1. A good electrical connection
2. A good mechanical connection

   The two metals to be joined are placed together and heat is applied to the joint. Solid solder is then placed on the heated joint. The solder becomes liquid and flows over the joint. When the heat is removed the solder cools and resolidifies (becomes solid again). The solder bonds the two metals, making a good mechanical connection and sealing the joint from outside air, thus preventing oxidation.

6-1a.2   Composition of Solder

There are different types of solder used in industry. Solder used for electronics is a metal alloy made up of combinations of tin and lead. Both of these metals are rather soft and have a low melting point compared to that of other metals. Pure lead has a melting point of 621 degrees Fahrenheit (°F), and pure tin melts at 450°F. When these metals are combined into a solder alloy of 60% tin to 40% lead (called a *60/40 solder*), the melting point of the alloy drops to 375°F.

The solder does not melt instantaneously, but rather, first becomes soft or plastic, then becomes liquid. Figure 6–1 illustrates the characteristics of a 60/40 solder. When the temperature of the solder is between room temperature and 361°F, it is in the solid condition. When the temperature rises above 361°F, the solder becomes plastic in nature. When the temperature reaches 375°F, the solder becomes liquid and will flow very easily. Similarly, as the solder cools it goes into the plastic condition before it resolidifies. For this reason, solder joints should not be subjected to any movement until they are completely solid, or a poor connection may result.

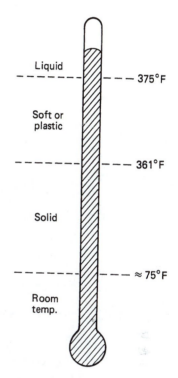

Figure 6–1 Temperature characteristics of 60/40 solder.

The most used type of solder in the electronics industry is the 60/40 type. Figure 6–2 shows a melting or fusion characteristic chart for other tin–lead combination solders. All other solder combinations have a plastic range, except the 63/37 type, which fully melts and solidifies at 361°F. The chemical symbol for tin is Sn, and this type of solder is called Sn 63. It is also known as *eutectic solder* and is more expensive to use.

A newer type of specialty solder, using *indium*, produces *indalloy solder*. Indium is a semiprecious, nonferrous, silvery-white metal with a brilliant luster. It is very malleable and ductile, being softer than lead.

Figure 6–2 Fusion characteristic of tin/lead solders. (Permission to reprint granted by PACE Incorporated, 9893 Brewers Court, Laurel, MD, 20707.)

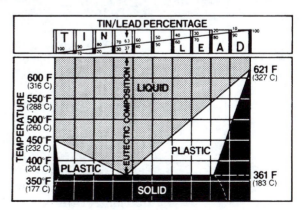

Because indium solder's melting point is less than that of conventional solder (indalloy No. 136 melts at 136°F) it is particularly well suited for soldering heat-sensitive electronic components.

6–1a.3 Wetting Action

Solder does more than just stick metal together in the form of a hot-metal glue. When hot solder comes into contact with a metal surface such as copper, a metal solvent action takes place, as shown in Figure 6–3. The *solvent action* results when hot solder dissolves and penetrates the copper surface. The molecules of the solder and copper combine in such a way as to produce highly bonded material with characteristics all its own. This solvent action, called *wetting*, forms the intermetallic bond between soldered parts.

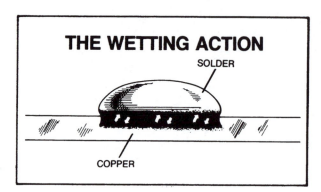

Figure 6–3 Molecular combination of solder. (Permission to reprint granted by PACE Incorporated, 9893 Brewers Court, Laurel, MD, 20707.)

6–1a.4 Flux and Acid

Surface oxides of metallic parts inhibit the soldering process even though they may appear to be clean. The very thin oxide film prevents the solder from reaching the metal and little or no wetting is accomplished. When hot solder comes in contact with the oxide film, it will act like a drop of water on an oily surface. There is no molecular bonding and the solder can be scraped away very easily. Cleaning the parts prior to soldering will help some, but because oxidation occurs so rapidly, the oxide film will quickly reappear. To overcome these oxide films, it is necessary to use materials called *fluxes* when soldering.

Flux is a soft substance of natural or synthetic rosins (perhaps with chemical additives called *activators*) which removes the thin film of oxide during the soldering process. The flux has a melting point just below that of solder. It is noncorrosive and nonconductive below this point. The flux can be placed on the joint prior to soldering. During the soldering process the flux melts first and vaporizes, which produces a corrosive action. This corrosive action removes oxides and prevents them from re-forming while the solder flows over the joint to form the desired intermetallic bond.

A wide variety of fluxes are used in many applications. For example, acid flux is used in soldering sheet metal, and borax paste is used in silver brazing. Usually, these fluxes require a higher melting temperature than that of the rosin flux used in electronic component soldering. Acid flux is very corrosive and can damage delicate electronic parts. *For electronic soldering use only rosin flux.*

To minimize the time required to place flux on a joint each time a solder connection is made, *rosin-core solder* was developed, shown in Figure 6–4. Rosin-core solder has a hollow path or hole in the center of the solder in which flux is placed. This provides a convenient way to apply and control the amount of flux used at a joint.

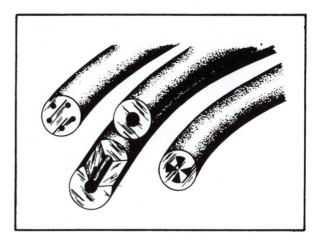

Figure 6–4 Rosin-core solder. (Permission to reprint granted by PACE Incorporated, 9893 Brewers Court, Laurel, MD, 20707.)

6–1b SOLDERING TOOLS AND METHODS

Like any tools, soldering tools must be used properly and safely to prevent injury to operating personnel and damage to materials. Particular attention must be used with hot soldering irons, which can burn, and with molten solder, which can splash or drip on the body and cause severe burns.

6–1b.1 Basic Soldering Iron

Heat must be used with any type of soldering. Soldering electronic parts usually involves the use of a conductive type of soldering iron, shown in Figure 6–5. Soldering irons come in a variety of shapes, sizes, and wattage ratings but consist of three basic elements: the handle (Figure 6–5a), which provides thermal insulation to the operator; the resistive heater unit (Figure 6–5b), which contains a coil of wire that passes electrical current for heating; and the tip (Figure 6–5c), which conducts the heat to the joint.

    The heater unit screws into the handle, and the tip screws into the heater unit. The assembly of these parts must be snug, but not overtight, which could strain the parts and cause damage or result in poor heating. This type of soldering iron can easily be held in the hand and uses about 50 watts of power.

6–1b.2 Basic Soldering Tools

To produce reliable solder connections and provide safety, other basic soldering tools are required. Figure 6–6 shows a sampling of these tools.

Figure 6–5 Basic soldering iron: (a) handle; (b) heater; (c) tip.

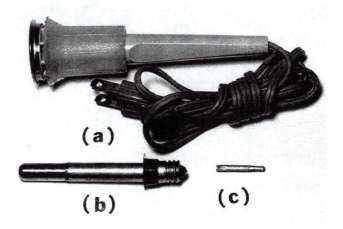

Flux (Figure 6–6a) is used to clean away the oxide film on component leads and connections. Rosin-core solder (Figure 6–6b), can be used in place of regular flux to speed up soldering time. Alcohol is used with a stiff bristle brush (Figure 6–6c) to clean away residue flux from a solder joint. The soldering iron holder (Figure 6–6d) provides a safe place to rest the iron between soldering operations. Soldering aids (Figure 6–6e) are plastic-handled devices with aluminum or stainless steel pointed tips and/or a small wire brush. Solder does not adhere to the tips very well, and these tools can be used to hold parts in place during a soldering operation or to remove excess solder or debris from a joint after it has cooled. A complete soldering iron is shown in Figure 6–6f. A wet sponge (Figure 6–6g) is used to wipe the tip of the iron or to remove excess solder and oxidation prior to soldering a connection.

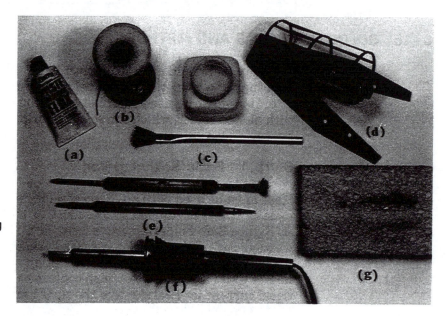

Figure 6–6 Basic soldering tools: (a) flux; (b) solder; (c) alcohol and brush; (d) soldering iron holder; (e) soldering aids; (f) soldering iron; (g) wet sponge.

6–1b.3 Proper Soldering Iron Preparation

Preparing a soldering iron for efficient soldering must be done properly, as shown in Figure 6–7. Initially, the iron must be plugged into an electrical outlet and allowed to heat to its full temperature (Figure 6–7a). Next, a small amount of solder is applied to the tip of the iron (Figure 6–7b). Both sides of the tip must be completely covered with molten solder. Finally, the tip is gently wiped on a moistened sponge (Figure 6–7c). This wiping action causes the molten solder to smooth out in a very even layer. Applying a small amount of solder to the tip of a soldering iron, wire, or component before the actual soldering operation begins is called *tinning*. An untinned soldering iron will oxidize quickly and make the soldering operation less efficient, because there is less heat transfer to the joint.

Figure 6–7 Soldering iron preparation (tinning): (a) heating; (b) applying solder or tinning; (c) wiping or cleaning on wet sponge.

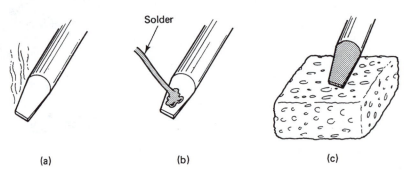

(a)　　　　　　　(b)　　　　　　　(c)

The area of contact between the soldering iron tip and the joint, called *thermal linkage,* is shown in Figure 6–8. An untinned soldering iron will have a small linkage area (Figure 6–8a). A tinned soldering iron will have a larger linkage area because of the slight solder bridge between the tip and the joint (Figure 6–8b). In some cases, more solder can be added between the tip and the joint to increase the solder bridge.

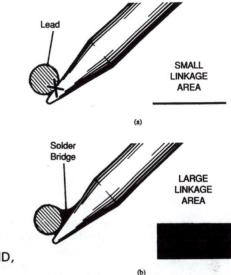

Figure 6–8 Thermal linkage: (a) small area without solder; (b) larger area with solder. (Permission to reprint granted by PACE Incorporated, 9893 Brewers Court, Laurel, MD, 20707.)

6–1b.4 General Soldering Method

Several special soldering methods are in use today, depending on the type of application. The general soldering method is used to connect wires and components to terminals, as shown in Figure 6–9.

The soldering iron must have been prepared as described earlier. The parts to be soldered must be stationary. In this case a tinned wire

Figure 6–9 General soldering procedures: (a) clean tip on wet sponge; (b) place iron tip on connection for short warm-up time; (c) apply solder to joint; after solder flows freely, remove solder and iron; (d) clean joint with brush and solvent (alcohol).

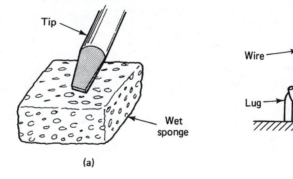

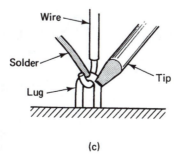

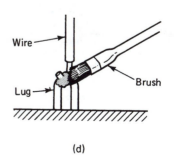

is bent around a standard solder lug terminal. First, the tip of the tinned iron is wiped on the wet sponge to remove any oxidation (Figure 6–9a). Next, the iron tip is placed on one side of the joint to heat it (Figure 6–9b). A slight amount of solder may be placed between the iron tip and the joint to produce a solder bridge. Solder is then placed on the other side of the heated joint (Figure 6–9c). The solder is kept on the joint until a reasonable amount of molten solder covers the joint. Then the solder is removed first, followed by the iron, and the joint is allowed to harden. A small brush dipped into a noncorrosive solvent or isoprophl alcohol is used on the joint to remove residual flux and debris (Figure 6–9d). The joint should appear smooth and shiny. Looking down from the top toward the connection, the solder should cover the wire slightly and feather out to a thin edge, as shown in Figure 6–10. This is referred to as proper wetting of a joint.

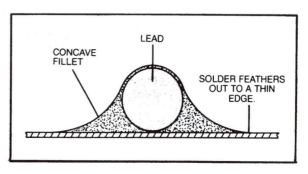

Figure 6–10 Proper wetting (round lead on flat surface). (Permission to reprint granted by PACE Incorporated, 9893 Brewers Court, Laurel, MD, 20707.)

6–1b.5 Types of Soldering Joints

The proper amount of solder and good wetting action determine a highly reliable solder joint. Figure 6–11 shows a correct solder joint and some

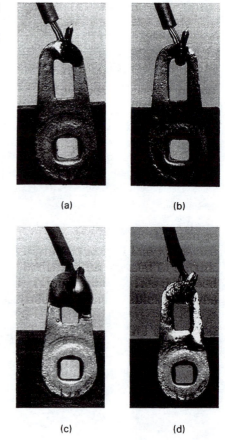

Figure 6–11 Solder joints: (a) correct; (b) too little solder; (c) too much solder; (d) fractured joint.

incorrect joints. A good solder joint (Figure 6–11a) has sufficient solder to cover the parts and connection. Too little solder on a joint (Figure 6–11b) may not withstand vibration and work itself loose. The solder may not cover the entire connection and has not formed a good electrical bond. This type of connection, referred to as a *cold solder joint*, usually results from insufficient heat of the iron or incorrect heating of the component lead and terminal lug. Too much solder on a joint (Figure 6–11c) is unnecessary and in some cases can short out with adjoining components. If the connection moves before the solder has a chance to cool and become hard, the connection will appear rough and gray in color. Such a connection, which may show very fine cracks in the solder, is called a *fractured joint* (Figure 6–11d).

Other unacceptable solder joints include the following:

Incomplete wetting, where insufficient solder has not totally covered the joint.

Excessive solder, where too much solder is placed on the joint. Remember, the outline of the wires at the joint must be visible.

Solder peaking, where small peaks of solder are seen at the joint as a result of less dwell time and/or insufficient heat. The solder did not flow completely over the joint.

Rosin joint, where insufficient heat results in a quantity of solidified flux between the connections of a joint.

Porous joint, where the solder has not properly wetted the joint and the cooling process produces holes in the solder of the joint.

Overheated joint, where the joint appears chalky, dull, or crystalline in appearance as a result of contamination of the joint or insufficient flux.

More information is given on solder joints in Unit 7.

6–1b.6   Types of Soldering Tips

A properly wetted solder joint depends on the correct amount of heat at the connection. The amount of heat transferred from the soldering iron to the connection is a function of the size of the iron tip. There are many types of soldering iron tips for specific uses. If too small a tip is used, insufficient heat may cause a cold solder joint. If too large a tip is used, the excess heat may damage components and materials. Figure 6–12 shows some types of soldering iron tips. A conical or needle tip is used

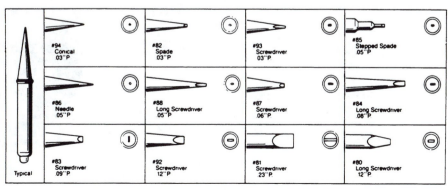

Figure 6–12 Soldering iron tips. (Permission to reprint granted by UNGAR, A Division of Eldon Industries, Inc., 5620 Knott Avenue, Buena Park, CA 90621.)

to solder thin wire and component lead holes in printed circuit boards. A screwdriver or spade tip is used to solder flat terminals.

6–1b.7 Temperature Control

The temperature of the soldering iron tip can also be controlled by electronic means. Figure 6–13 shows a basic soldering unit that has temperature control. The proper heat is selected by the control on the unit. This allows the operator to adjust the temperature lower for more sensitive circuits without changing the size of the soldering iron.

Controlling the temperature of the iron tip is not the main problem in soldering. The main consideration in temperature control in soldering is the *heat cycle:* the time required for the material (or workpiece) to become hot, the temperature of the material, and the period during which it remains hot. An ideal heat cycle would be a very fast heat-up time, sufficient heat to melt the solder, and a short period of heat after the iron is removed. This heat cycle is affected by factors other than the temperature of the soldering iron tip.

One factor to be considered is the size of the joint to be soldered. A large joint such as a component terminal will take longer to heat than will a smaller connection. The relationship of the size or mass of the workpiece to the temperature required to solder it is referred to as its *relative thermal mass.*

If the workpiece is mounted to a heat-conducting surface, the relative thermal mass increases and more heat is required for soldering. The soldering iron generates and stores the heat in its tip. When the tip is placed on the work, heat is transferred to the workpiece and the tip loses some of its stored heat. The heating element replenishes heat to the tip. If the relative thermal mass of the workpiece is so large that a lot of heat is lost by the tip, it may not be able to recover fast enough to bring the workpiece up to the correct temperature for soldering.

Another important factor in temperature control is the surface condition of the workpiece. Very little heat can be transferred to the workpiece from the tip of the iron if oxides or other contaminants are present. Before soldering a joint the workpiece should be cleaned with a solvent such as trichloroethane or isopropyl alcohol to remove any grease or oil film on the surface. Next, a fine abrasive material such as sandpaper can be used to remove existing oxides. Remember to remove only the thin oxide film, not too much of the workpiece material. In some cases, a pencil eraser can be used to clean a workpiece.

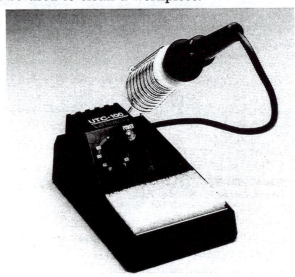

Figure 6–13 Basic soldering unit with holder temperature control and wet sponge. (Permission to reprint granted by UNGAR, A Division of Eldon Industries, Inc., 5620 Knott Avenue, Buena Park, CA 90621.)

Temperature control can also be maintained by increasing the thermal linkage as discussed in Section 6–1b.3 and shown in Figure 6–8. By adding a little solder to the tip of the iron the solder bridge becomes larger and more heat can be transferred to the workpiece.

The best way to determine if you have the correct temperature control for soldering is to observe the rate at which solder melts, referred to as *heat rate recognition*. An ideal rate is 1 to 2 seconds. For proper soldering, the time an iron is left on the workpiece, called the *dwell time*, should not be longer than 1 to 2 seconds.

6–1b.8 Tinning a Wire

Very often it is necessary to seal the ends of stranded wire to be soldered in cables and other electronic circuits. First the insulation must be removed from the wire as shown in Section 4–1a.10. Remember not to nick or cut any of the strands of wire and to trim the insulation away very neatly.

The strands of the stripped wire must be close together and form a round appearance. They can be formed with the fingers in a rotating manner, if needed. The solder must cover the end of the wire but must not touch the insulation. A heat sink of some type can be placed next to the insulation, which will prevent the solder from moving into the insulation. Long nose pliers, alligator clip leads, or special devices can be used as heat sinks. The special antiwicking tool shown in Figure 6–14 is easier to use than pliers. The solder is allowed to flow over the stripped wire from the antiwicking tool to the end of the wire, as shown by the arrows. Remember not to use too much solder on the wire.

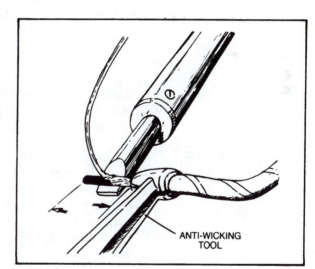

Figure 6–14 Tinning a wire. (Permission to reprint granted by PACE Incorporated, 9893 Brewers Court, Laurel, MD, 20707.)

ANTI-WICKING TOOL

6–1b.9 Splicing a Wire

Sometimes it is necessary to connect or *splice* two wires together for a long run or to repair a circuit. The wires must be stripped of insulation sufficiently to allow a good mechanical connection to be produced. The bare wires must come in contact with each other in some manner, as shown in Figure 6–15.

A simple hook splice (Figure 6–15a) may be all that is needed for some connections. Each wire has a hook formed in the stripped wire. The hook ends are placed together and sometimes crimped tight. A twisting and bending method (Figure 6–15b) may be required for more strength

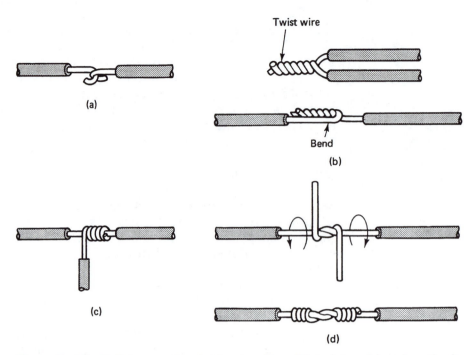

Figure 6–15 Splicing a wire: (a) hook splice; (b) twist and bend method; (c) wire junction (T-splice); (d) Western Union splice.

in the connection. Very often it is required to tap into one wire with another wire, using a wraparound method, as shown in Figure 6–15c. A segment of the insulation must be removed carefully from the main wire. The stripped end of the other wire is then wrapped around the bare segment of wire. This type of connection is referred to as a *T-splice*.

The *Western Union splice* (Figure 6–15d) provides a good mechanical connection with the least amount of bulk. The two stripped wires are wound around each other in an opposite manner. Then the ends are trimmed so that no wire is protruding above the insulation. For all types of splices the wires are soldered together to form a good mechanical and electrical connection. The important thing to remember is to keep the splice as small as possible, preferably less than the diameter of the outer insulation.

An insulated covering must be placed around the spliced wire and should not be bulky, since this may interfere with other circuitry. Very often, especially if it is a repair job, black electrical tape is used for insulation to wrap around the splice.

Heat-shrinkable tubing is plastic tubing that contracts or shrinks with the application of heat. Heat-shrinkable tubing provides excellent insulation for a spliced wire. A heat gun similar to a hand-held hair dryer is used to shrink the tubing. A heat gun is shown in Figure 6–16.

Figure 6–16 Heat gun. (Permission to reprint granted by UNGAR, A Division of Eldon Industries, Inc., 5620 Knott Avenue, Buena Park, CA 90621.)

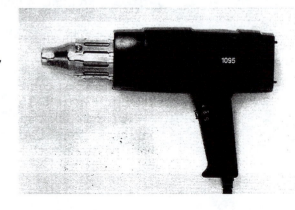

Chap. 6 / Soldering Techniques

Heat-shrinkable tubing is cut long enough to cover the splice and a little of the insulation on each wire. This tubing is slid out of the way during the splicing and soldering of the wire, as shown in Figure 6–17a. Once the splice has cooled, the tubing is slid over the splice and heat is applied as shown in Figure 6–17b. Too much heat must not be applied or the tubing may burn and split open. The tubing shrinks and forms a tight-fitting insulated covering around the splice, as shown in Figure 6–17c

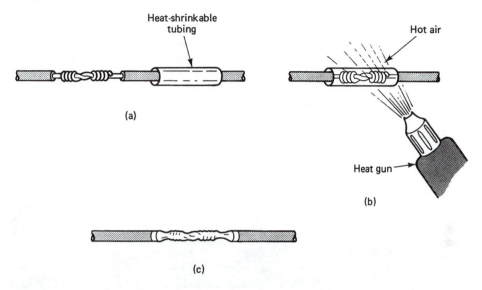

(a)

(b)

(c)

Figure 6–17 Use of heat-shrinkable tubing: (a) splice with tubing slid back; (b) tubing over splice and heat gun applied; (c) tubing shrinks to contour of splice.

6–1b.10 Removing Solder

Sometimes it is necessary to remove hardened solder from an unacceptable joint to be able to resolder it correctly. Also, solder must be removed from components when repairing electronic equipment. Several devices can be used effectively to remove solder from a joint or component, as shown in Figure 6–18. A *wicking braid* (Figure 6–18a) consists of stranded–braided bare copper wire saturated with flux which is used to draw molten solder into it by capillary action. A *desoldering bulb* (Figure 6–18b) is a rubber bulb with a small nylon nozzle. The bulb is

Figure 6–18 Desoldering tools: (a) wicking braid; (b) desoldering "vacuum" bulb; (c) spring-loaded desoldering tool; (d) desoldering iron with vacuum bulb.

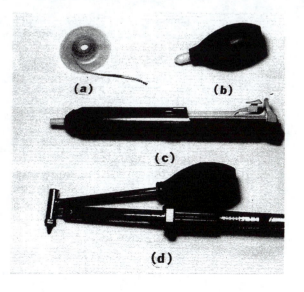

(a)

(b)

(c)

(d)

depressed with one hand and then placed on molten solder. The bulb is released and the solder is drawn up into the nozzle with a vacuum action. The cooled solder can be ejected from the bulb later.

A *spring-loaded desoldering tool* (Figure 6–18c) also provides vacuum action. The lever is pushed down until it locks in place. The tool nozzle is placed on the molten solder and then the lever is released, which draws the solder up into the tool. A combination soldering iron and bulb is also available as a desoldering tool (Figure 6–18d). The desoldering tool is best used on removing semiconductor devices and ICs. The action of this type of desoldering tool is known as *vacuum pulse desoldering,* since a quick vacuum pulse removes solder.

Removing solder with a wicking braid requires a couple of basic steps, as illustrated in Figure 6–19a and b. The braid to be used to remove solder must be clean. Used braid must be cut away with a pair of diagonal cutters. Even though the wicking braid comes with flux, it is better to place a little more flux on it.

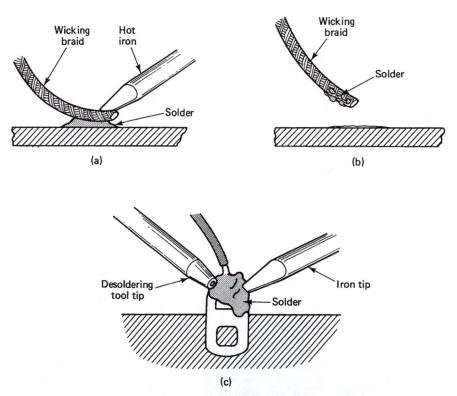

Figure 6–19 Solder removal: (a) wicking braid and soldering iron; (b) solder drawn into braid by capillary action; (c) a vacuum pulse desoldering tool and soldering iron.

Safety Hint! Heat can travel up the braid and burn your hand. Hold the braid with the plastic or nylon container that it comes in. This will prevent your hand from getting burned.

The braid is placed on the solder to be removed and then a hot soldering iron is placed on top of the braid (Figure 6–19a). The heat from the iron penetrates the braid and melts the solder. The molten solder is attracted into the braid by capillary action. The braid, now containing most of the solder, is then removed (Figure 6–19b).

Capillary action is the attraction of a liquid to a solid. Water droplets on a glass window pane or automobile after a rain storm are evidence of capillary action. Heat causes solid solder to melt. The molten or liquid solder tends to flow toward the heat. The molecules of molten solder will be attracted to a well-heated surface of metal rather than being attracted to each other within the solder. Just as a paper towel soaks up water, a wicking braid soaks up molten solder.

With a desoldering tool, the soldering iron is placed on one side of the joint, as shown in Figure 6–19c. When the solder becomes molten, the desoldering tool nozzle is placed on the joint and a vacuum is created that draws the solder into the nozzle.

6–1c SOLDERING TERMINALS

There are many types of terminals used for connecting wires and components in electronic assembly. Each type of terminal usually requires a modification of the general soldering procedure. Wires must be stripped and pretinned before soldering to terminals as described in Section 6–1b.8. In this section we describe the soldering procedures for three common types of terminals.

6–1c.1 Turret Terminal

The *turret terminal*, shown in Figure 6–20, is round. The wire is wrapped around it using long nose pliers. The distance that the wire is wrapped varies from one-half turn (180°) to three-fourths turn (270°) or a full turn (360°). One-half turn does not provide the best position against movement while soldering, but the wire can be removed easily. A full-turn wrap gives the best grip, but the three-fourths turn wrap is also in common use, depending on the type of job desired. There should be a distance of one to two diameters of wire between the wire insulation and the entry point of the terminal. Several wires can be soldered to this terminal, but only one wire connection will be demonstrated.

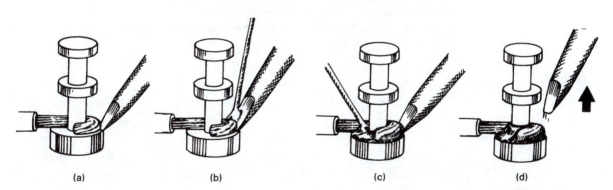

(a) (b) (c) (d)

Figure 6–20 Soldering a turret terminal: (a) applying iron to point of maximum thermal mass; (b) making a solder bridge to increase thermal linkage; (c) applying solder to side opposite the iron; (d) removing iron tip with a forward wiping motion. (Permission to reprint granted by PACE Incorporated, 9893 Brewers Court, Laurel, MD, 20707.)

There are four steps in soldering a turret terminal:

1. Apply an iron to the maximum thermal mass.
2. Melt a small amount of solder on the iron to form a solder bridge.
3. Apply solder to the side opposite the iron. Allow the solder to melt and flow over the joint.
4. Remove the solder, and then the iron, with a forward wiping motion.

After the joint has solidified and cooled, clean it with a solvent such as alcohol and a small stiff brush. Each solder joint must be inspected to ensure a reliable connection. Figure 6–21 shows a finished soldered turret terminal.

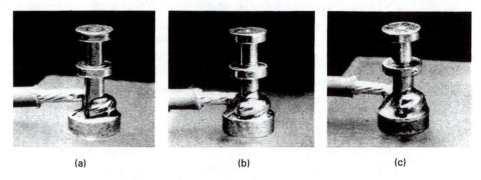

(a)   (b)   (c)

Figure 6–21 Soldered turret terminal: (a) minimum acceptable solder; (b) preferred joint; (c) maximum acceptable solder. (Permission to reprint granted by PACE Incorporated, 9893 Brewers Court, Laurel, MD, 20707.)

The most preferred joint should have sufficient solder and appears bright and shiny, as shown in Figure 6–21b. You should be able to see the individual strands of wire, and the solder should feather out smoothly, indicating good wetting action along all elements of the joint. Figure 6–21a and c show the minimum and maximum acceptable solder, respectively, for a turret terminal.

6–1c.2 Bifurcated Terminal

The *bifurcated terminal* has two upright posts with a hole through the terminal, as shown in Figure 6–22. Wires can enter the terminal from the side, top, or bottom. Normally, the wire is placed through the posts and bent at a right angle (90°) with a pair of long nose pliers to form a mechanical connection. The wire is cut flush with the terminal. The insulation of the wire should also be one to two diameters from the point of entry of the terminal.

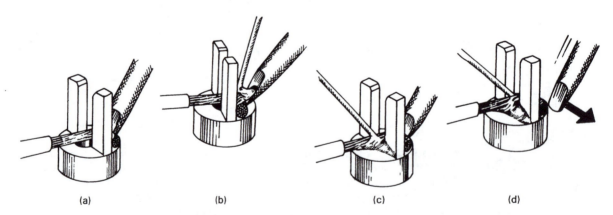

(a)   (b)   (c)   (d)

Figure 6–22 Soldering a bifurcated terminal: (a) applying iron to point of maximum thermal mass; (b) making a solder bridge to increase thermal linkage; (c) applying solder to side opposite the iron; (d) removing iron with wiping motion. (Permission to reprint granted by PACE Incorporated, 9893 Brewers Court, Laurel, MD, 20707.)

The soldering procedure for a bifurcated terminal is similar to that for the turret terminal:

1. Apply an iron to the maximum thermal mass.
2. Melt a small amount of solder on the iron to form a solder bridge.
3. Apply solder to the side opposite the iron. Allow the solder to melt and flow over the joint.

4. The iron is removed first with a side wiping motion, and then the solder is removed before it solidifies.

The joint is cleaned and inspected as with other soldering procedures. Figure 6–23 shows acceptable finished solder joints for a bifurcated terminal.

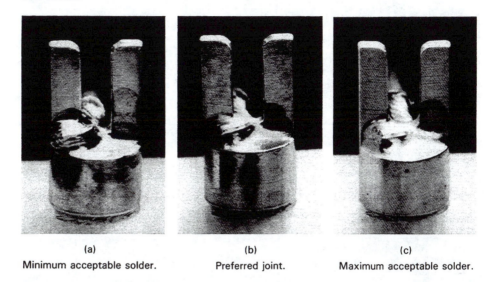

(a)	(b)	(c)
Minimum acceptable solder.	Preferred joint.	Maximum acceptable solder.

Figure 6–23 Soldered bifurcated terminal: (a) minimum acceptable solder; (b) preferred joint; (c) maximum acceptable solder. (Permission to reprint granted by PACE Incorporated, 9893 Brewers Court, Laurel, MD, 20707.)

6–1c.3 Cup Terminal

Cup terminals may be placed on circuit boards at various locations, but they are commonly grouped together in cable connectors (see Figure 1–8c). A cup terminal is actually a hollow cylinder into which wire is soldered without any additional mechanical support. A specific amount of solder is melted in the cup and then the wire is inserted.

The procedure for soldering a cup terminal (Figure 6–24) is as follows:

1. Trim the pretinned wire to the correct length to fit in the cup. Remember to leave one to two diameters of bare wire before the insulation.
2. Twist a small length of solder together so that it has a loop at one end.
3. Place the twisted solder into the cup and cut it the length of the cup.
4. Place an iron on the back of the cup. Leave the iron there until the solder begins to melt. Now insert the wire into the cup. Continue to leave the iron on the cup until the flux boils or bubbles (not vaporized) to the surface. The iron is now removed, but the wire must be held stationary until the solder solidifies.

Figure 6–25 shows some acceptable solder joints of a cup terminal.

6–1d THE SOLDERING GUN

For noncritical circuits and in repair work, a soldering gun is used as shown in Figure 6–26. The soldering gun has a step-down transformer which connects to a heating element and copper tip. The gun can be left plugged into an ac electrical outlet continuously, but will not be on. Once

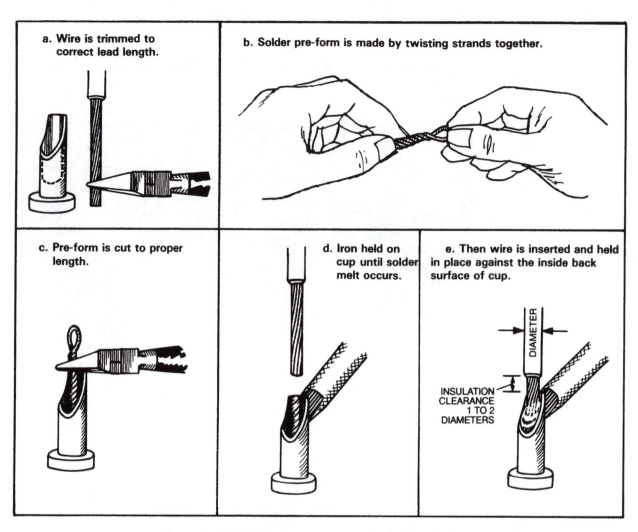

a. Wire is trimmed to correct lead length.

b. Solder pre-form is made by twisting strands together.

c. Pre-form is cut to proper length.

d. Iron held on cup until solder melt occurs.

e. Then wire is inserted and held in place against the inside back surface of cup.

DIAMETER

INSULATION CLEARANCE 1 TO 2 DIAMETERS

Figure 6–24 Soldering a cup terminal. (Permission to reprint granted by PACE Incorporated, 9893 Brewers Court, Laurel, MD, 20707.)

Figure 6–25 Soldered cup terminal: (a) minimum acceptable solder; (b) preferred joint; (c) maximum acceptable solder. (Permission to reprint granted by PACE Incorporated, 9893 Brewers Court, Laurel, MD, 20707.)

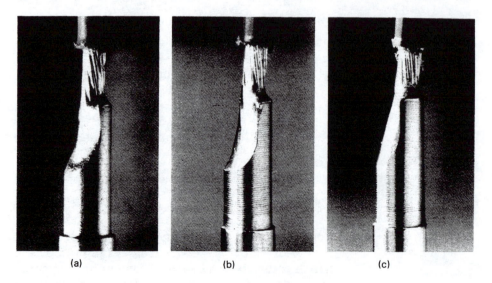

(a)   (b)   (c)

Figure 6–26 Soldering gun. (Permission to reprint granted by UNGAR, A Division of Eldon Industries, Inc. 5620 Knott Avenue, Buena Park, CA 90621.)

the trigger is activated, voltage is applied to the gun and it heats up very rapidly. The gun's heat is not controllable, and the ac voltage produces electromagnetic radiation that can destroy metal-oxide-semiconductor (MOS) solid-state devices.

SECTION 6–2
DEFINITION EXERCISES

Write a brief description of each of the following terms.

1. Soldering _____

_____

2. Solder _____

_____

3. Oxide _____

_____

4. Eutectic solder _____

_____

5. Solvent action _____

_____

6. Wetting _____

_____

7. Flux _____

_____

8. Rosin-core solder _____

_____

9. Indium _____

_____

10. Soldering iron _____

_____

11. Soldering aid _____

_____

12. Soldering iron holder _____

_____

13. Tinning _____

_____

14. Thermal linkage _____

_____

15. Cold solder joint _____

_____

16. Fractured solder joint _____

_____

17. Incomplete wetting _____

_____

18. Excessive solder _____

_____

19. Solder peaking _____

_____

20. Rosin joint _____

_____

21. Porous joint _____

_____

22. Overheated joint _____

_____

23. Temperature control _____

_____

24. Heat cycle _____

_____

25. Relative thermal mass _____

_____

26. Heat recognition time _____

_____

27. Dwell time _____

_____

28. Antiwicking tool _____

_____

29. Splice _____

_____

30. Heat-shrinkable tubing _____

_____

31. Wicking braid _____

_____

32. Desoldering bulb _____

_____

33. Desoldering iron _____

_____

34. Spring-loaded desoldering tool _____

_____

35. Capillary action _____

_____

36. Soldering gun _____

_____

37. Solder sucker _____

_____

38. Vacuum pulse desoldering _____

_____

≡≡≡ **SECTION 6-3**
      EXERCISES AND PROBLEMS

Complete this section before beginning the next section.

1. List the two main uses of soldering.

   a.

   b.

2. List at what temperature the following metals or combination metals become liquid.

   a. Pure lead   ____ °F

   b. Pure tin   ____ °F

   c. 60/40 solder   ____ °F

   d. Eutectic solder   ____ °F

   e. Indalloy No. 136   ____ °F

3. Why is flux used in the soldering process?

4. Refer to Figure 6–5 and list the three basic parts of a soldering iron.

   a.

   b.

   c.

5. Refer to Figure 6–7 and list the three steps for tinning and soldering iron preparation.

   a.

   b.

   c.

6. Refer to Figure 6–9 and explain briefly the four general soldering procedures.

   a.

   b.

c.

d.

7. Refer to Figure 6–11 and list three solder joint connections that are unacceptable.

a.

b.

c.

8. Identify where the following soldering iron tips can be used.

a. Conical

b. Screwdriver

c. Needle

d. Spade

9. List the steps in removing solder using wicking braid.

a.

b.

c.

d.

10. Describe the safety hint when using wicking braid.

11. List the three parts of the heat cycle.

a.

b.

c.

12. List the principal factors affecting temperature control during the soldering process.

a.

b.

c.

d.

≡ SECTION 6–4
EXPERIMENTS

EXPERIMENT 1. Stripping and Tinning a Wire

Objective:

To develop the skills needed to properly strip and tin an insulated stranded wire.

Introduction:

In nearly all cases, stranded wires must be stripped and tinned before they are placed in circuits. You must be very careful when stripping wire that you do not cut or nick any of the strands. Missing or nicked strands cannot carry the full amount of current designed for a circuit and the mechanical connection is also weakened. It is also important that a proper solder joint be made for the reliability of the circuit.

Materials Needed:

1. Wire stripper
1 Pair of diagonal cutters
1 Heat sink clip or antiwicking tool
1 50-watt (maximum) soldering iron
1 Soldering iron holder
1 Moist sponge
1 Container of flux
1 Small soldering vise or project holder
   Some 60/40 rosin-core solder (No. 22)
   Some 22-gage insulated wire

Procedure:

1. Plug in the soldering iron and place it in the holder
2. After the iron is hot, prepare the iron according to the information given in Section 6–1b.3.
3. Adjust the wire stripper as shown in Figure 4–16a, to remove the insulation of the wire but not cut or nick the wire.
4. Strip about $\frac{1}{2}$ inch of the insulation from the wire as shown in Figure 4–16b.
5. Inspect the wire for cut or nicked strands. If you find any, cut the bare wire off with the diagonal cutters and repeat steps 3 and 4.
6. Place the wire in the vise on the insulation but do not clamp it so hard as to distort the wire.
7. Clip the heat sink clip on the bare wire next to the insulation or use the antiwicking tool as shown in Figure 6–14.
8. Make sure that all the strands of wire are in place.
9. Place the iron on the bare wire and add a little solder as explained in Section 6–1b.8.
10. Let the solder solidify and cool down before you inspect the tinned wire. Solder should only be on the last half of the bare wire and not have reached the insulation. The insulation should be neat and even around the wire.

Fill-in Questions:

1. When stripping wire you must not

_____ or _____ the strands of wire.

2. Flux is used to clean away _____

_____ on a joint just before soldering.

3. Before beginning any soldering job the soldering iron must be _____ .

4. A properly tinned wire does not have solder touching the _____ .

EXPERIMENT 2. Splicing Wires

Objective:

To gain skill in joining two wires together in the form of a splice.

Introduction:

In this experiment you will perform the four types of wire splices shown in Figure 6–15. Single-strand wire is the easiest to splice; however, stranded wire can be used instead. After the insulation is removed from the wire, the strands must be twisted together tightly in an attempt to form a single wire. Some tinning of the wires can be used, but tinned wire is much more difficult to twist into a splice.

Materials Needed:

1 Wire stripper
1 Pair of diagonal cutters
1 Pair of long nose pliers
1 Heat sink clip or antiwicking tool
1 50-watt (maximum) soldering iron
1 Soldering iron holder
1 Moist sponge
1 Container of flux
1 Small soldering vise or project holder
   Some 60/40 rosin-core solder (No. 22)
   Some single-strand insulated wire, 18 to 22 gauge
   Some 22-gage insulated wire

Procedure:

1. Prepare the soldering iron as described in Experiment 1.
   Hook Splice

2. Refer to Figure 6–15a and remove about $\frac{1}{2}$ in. of insulation from the end of each wire.

3. Using the long nose pliers, form a small hook on the end of each of the stripped wires.

4. Place the hook ends together and lightly crimp with the long nose pliers.

5. Solder the connection.

6. Refer to Figure 6–15b and strip about $\frac{3}{4}$ in. of insulation from the wires.
   Twist and Bend Method

7. Twist the wires together.

8. Fold back one of the wires.

9. Solder the connection.

10. Refer to Figure 6–15c and strip about $\frac{1}{2}$ in. of insulation from one wire.
    T-splice

11. Carefully remove about $\frac{1}{4}$ in. of insulation from the center of the other wire. Do not nick the wire, and trim any excess insulation from wire. The soldering iron can be used to melt the insulation for easier removal.

12. Wrap the end of the one wire around the other wire as shown. Keep the spacing between turns of wire as close as possible. Cut off the tip of the spliced wire so that it is below the insulation.

13. Solder the connection.

14. Refer to Figure 6–15d and strip about 1 in. of insulation from the end of each wire.
    Western Union Splice

15. Wrap the two wires in opposite directions around each other.

16. Trim the ends of the wires so that they are below the insulation.

17. Solder the connection.

Fill-in Questions:

1. When splicing two wires, you must be sure to have a good _____ and _____ connection.

2. The turns or wraps of wire in a splice must be as _____ as possible.

3. It is preferable to have the size of a splice _____ than the diameter of the outer insulation.

EXPERIMENT 3. Soldering Terminals

Objective:

To demonstrate the skills required to solder various terminals properly.

Introduction:

In this experiment you will perform soldering procedures on four types of terminals.

Materials Needed:

1 Wire stripper
1 Pair of diagonal cutters
1 Pair of long nose pliers
1 Heat sink clip or antiwicking tool
1 50-watt (maximum) soldering iron
1 Soldering iron holder
1 Moist sponge
1 Container of flux
1 Container of solvent or alcohol
1 Small stiff bristle brush
1 Small soldering vise or project holder
1 Standard solder lug terminal
1 Turret terminal
1 Bifurcated terminal
1 Cup terminal or cable connector
  Some 60/40 rosin-core solder (No. 22)
  Some 22-gage insulated wire

Procedures

1. Prepare the soldering iron.

2. Strip and tin four wires about 1 ft long.

3. Refer to Section 6–1b.4 and Figure 6–9 to solder a standard solder lug terminal.

4. Inspect your work and refer to Figure 6–11.

5. Refer to Section 6–1c.1 and Figure 6–20 to solder a turret terminal.

6. Inspect your work and refer to Figure 6–21.

7. Refer to Section 6–1c.2 and Figure 6–22 to solder a bifurcated terminal.

8. Inspect your work and refer to Figure 6–23.

9. Refer to Section 6–1c.3 and Figure 6–24 to solder a cup terminal.

10. Inspect your work and refer to Figure 6–25.

Fill-in Questions:

1. The wires soldered to terminals should

   be _____ .

2. The soldering iron should be wiped on

   the moist _____ before soldering each joint.

3. After each joint is soldered, you should

   clean it with a _____ or _____

   _____ .

4. The last step in soldering a joint is to

   _____ it.

5. Three types of unacceptable joints are

   _____ , _____ , and __

   _____ .

EXPERIMENT 4. Removing Solder from a Solder Joint

Objective:

To show how to remove solder from a terminal with a wicking braid and desol-dering tool.

Introduction:

There are several ways to remove solder from a joint. In the first part of this experiment you will use a wicking braid and the normal soldering tools. In the second part you will use a desoldering tool.

Materials Needed:

1 Wire stripper
1 Pair of diagonal cutters
1 Heat sink clip or antiwicking tool
1 50-watt (maximum) soldering iron
1 Soldering iron holder
1 Desoldering tool (rubber bulb or spring-loaded type)
1 Moist sponge
1 Container of flux
1 Small soldering vise or project holder
1 Roll of wicking braid
1 Presoldered joint or small patch of solder on a terminal
  Some 60/40 rosin-core solder (No. 22)
  Some 22-gage insulated wire

Procedure:

1. Prepare the soldering iron.
2. Refer to Section 6–1b.10 and Figure 6–19 to remove solder with a wicking braid.
3. Cut off any used part of the wicking braid with a pair of diagonal cutters.
4. Dip the braid into the container of flux.
5. Place the braid on the solder joint.
6. Place the soldering iron on top of the braid.
7. Allow the joint to heat up until you see some of the solder rising through the top of the braid and around the iron tip.
8. Continue to hold the iron in place a couple of seconds more in order to draw as much solder as possible into the braid.
9. Remove the iron and the braid from the joint.
10. Notice how much solder is on the braid and look to see how much solder is left on the joint. Steps 3 to 10 can be repeated to remove any remaining solder if it is needed.

Use a desoldering tool to remove solder in the following procedures.

11. Set the desoldering tool if it is of the spring-loaded type.
12. Place the iron on the joint with one hand.
13. With the other hand, grip the desol-dering tool (depress the rubber bulb if it is this type).
14. When the solder becomes molten, remove the iron and quickly place the desoldering tool on the joint. Imme-diately release the bulb (or spring) to cause the vacuum pulse action.
15. Steps 11 to 14 can be performed several times to remove as much solder as possible from the joint.
16. The iron is again placed on the joint and the wire is gently shaken off the terminal.

Fill-in Questions:

1. Any used wicking braid should be

   _____ before trying to use it on
   solder removal from a joint.

2. For efficient solder removal, you should

   dip the wicking braid into _____
   before placing it on the joint.

3. The molten solder is drawn into the

   wicking braid by _____ action.

4. The molten solder is drawn into the

   solder sucker by _____ action.

5. Manual desoldering tools are referred to

   as constituting the _____
   vacuum method of desoldering.

≡ SECTION 6–5
INSTANT REVIEW

- *Soldering* is done for two reasons: to provide (1) a good electrical connection and (2) a good mechanical connection.
- *Solder* is a metal alloy of tin and lead.
- Solder generally used for electronic work is called *60/40* which means that it is 60% tin and 40% lead.
- *Eutectic solder* is solder that has no plastic state and becomes liquid and soldifies at exactly 361°F.
- *Oxide* is a chemical compound that forms on metal surfaces when the surfaces are exposed to air. Oxides prevent a good solder connection.
- *Solvent action* is the washing away or dissolving of oxides on a metal surface.
- *Wetting action* occurs when molecules of liquid solder combine with molecules of a metal surface and form a bond of a slightly different material.
- *Flux* is a soft rosin-type substance with a melting temperature just below that of solder, used to dissolve the oxides an instant before soldering is accomplished.
- *Rosin-core solder* is solder with a small tunnel of rosin in its center.
- *Indium* is a soft nonferrous metal used in solder because its melting point, 136°F, is less than that of normal solder.
- A *soldering iron* is a tool used for soldering that consists of a handle, a heating element, and a tip.
- A *soldering aid* is a plastic-handled device with aluminum or stainless steel tips used to remove rosin, unwanted solder bits, and other debris from solder joints.
- A *soldering iron holder* is a device used to rest and protect the soldering iron while it is not being used.
- *Relative thermal mass* refers to the size of the joint to be soldered.
- The *size* of a soldering iron and tip depend on the type of soldering job. A larger iron and special tip are needed for a large thermal mass.
- *Conical* and *needle tips* are used to solder thin wire and component leads.
- *Screwdriver* and *spade tips* are used to solder flat terminals.
- *Tinning* is the process of adding a small amount of solder to the iron tip or wires before the soldering joint is made.
- *Thermal linkage* is the heat transfer between the iron and the workpiece or joint.

- A *solder bridge* is a small amount of solder placed between the tip of the iron and the workpiece to increase thermal linkage.
- *Soldering iron preparation* involves heating the iron, applying a small amount of solder to the tip for tinning, and wiping the tip on a moist sponge.
- The *general soldering procedures* are:
  1. Clean the iron tip on a moist sponge.
  2. Place the iron tip on the joint and allow a short warm-up period.
  3. Apply solder to the opposite side of the joint. When the solder melts and covers the joint sufficiently, remove the solder and the iron with an upward wiping motion.
  4. Clean the joint with solvent or alcohol.
- A *cold solder joint,* a joint with insufficient solder covering the entire joint, usually provides a poor electrical connection.
- A *fractured solder joint,* a joint that has moved before it had solidified, is usually a poor electrical and mechanical connection. It may appear white in color.
- *Temperature control* means controlling the temperature of the soldering iron tip and at the joint for efficient soldering.
- A good *heat cycle* involves a fast warm-up time, sufficient heat to melt the solder, and a short duration of heat after the iron is removed.
- *Heat recognition time* refers to the actual time it takes for solder to melt at a joint. This time should be from 1 to 2 seconds but no longer, so as not to burn work materials.
- *Dwell time* is the time the iron is left on the workpiece.
- An *antiwicking tool,* used for tinning wires, prevents solder from running up under the insulation of a wire.
- A *splice* is the joining of two wires by mechanical and electrical means.
- *Heat-shrinkable tubing* is plastic tubing that can be placed on splices or terminals for insulation purposes. Heat applied to the tubing causes it to shrink and adhere to the physical contours beneath it.
- A *wicking braid* consists of stranded–braided bare copper wire saturated with flux that is used to draw molten solder into it by capillary action.
- *Capillary action* is the attraction of a liquid to a solid.
- A *desoldering bulb* is a small rubber bulb with a nylon tip used to remove molten solder from a joint by creating a vacuum when it is depressed and released.
- A *desoldering tool* is a soldering iron and desoldering bulb built into the same device.
- A *spring-loaded desoldering tool* uses a return spring to create a vacuum for drawing away molten solder.
- Various *solder terminals* are used for different applications. Although the soldering procedures are similar, wires are joined at each terminal in a specific manner. Terminals include the standard solder lug type, turret, bifurcated, and cup.
- A *soldering gun* is a device that heats up very fast because of the transformer it uses. A soldering gun is normally off until its trigger is activated. It is used for some limited assembly work and repairs. The electromagnetic field around the gun when it is on can destroy MOS solid-state devices.

Circle the most correct answer for each question.

1. Solder consists of:

   a. tin and copper   **b.** tin and lead

   c. lead and copper   **d.** lead and silver

2. Solder generally used for electronic work is rated:

   a. 63/37   **b.** 50/50

   c. 40/60   **d.** 60/40

3. A solder that melts and solidifies at 361°F is called:

   a. indium alloy No. 136   **b.** pure tin

   c. pure lead   **d.** eutectic

4. The combining of two metals into a bond of slightly different material is called:

   a. solvent action   **b.** wetting action

   c. capillary action   **d.** none of the above

5. A substance that removes oxides from the surface of metals is called:

   a. alcohol   **b.** flux

   c. both of the above   **d.** none of the above

6. Normally, soldering a joint involves:

   a. placing the solder on the iron

   b. placing the solder on the heated workpiece

   c. placing the solder on the insulation

   d. none of the above

7. Just before you solder a joint you should:

   a. place solder on the soldering iron

   b. place solder on the workpiece

   c. wipe the soldering iron on a moist sponge

   d. clean the soldering tip with alcohol

8. A solder bridge is used to:

   a. increase thermal linkage

   b. remove excess solder

   c. remove used flux

   d. support the weight of a stranded wire

9. Dwell time is the:

   a. time it takes to solder a joint completely

   b. time it takes to tin a wire

   c. actual time the soldering iron is on the joint

   d. time used to test an automobile

10. One size of soldering iron is efficient for all jobs.

    a. True   **b.** False

11. The most important element in temperature control of the soldering process is the temperature of the soldering iron tip.

    a. True   **b.** False

12. A large thermal mass is difficult to solder with a small soldering iron tip.

    a. True   **b.** False

13. A wicking braid is used to:

    a. remove solder

    b. prevent solder from flowing into the insulation of a wire

    c. strengthen a solder joint

    d. none of the above

14. The best soldering iron tip to use for soldering flat terminals is the:

    a. conical   **b.** needle

    c. spade   **d.** all of the above

15. Unacceptable solder joints will have:

    a. too little solder   **b.** too much solder

    c. a whitish look   **d.** all of the above

16. When soldering wires, the first thing to do is:

   a. place the wire on the terminal

   b. place solder on the terminal

   c. tin the wire

   d. clean the wire in a solvent

17. The best insulation to place over a wire splice is:

   a. masking tape

   b. plastic electrical tape

   c. heat-shrinkable tubing

   d. none of the above

18. The phenomenon that causes solder to move toward a heated surface of metal is called:

   a. wetting action   **b.** capillary action

   c. solvent action   **d.** magnetic attraction

19. A tool that uses a vacuum bulb is called:

   a. an antiwicking tool   **b.** a desoldering tool

   c. a soldering aid   **d.** a wicking braid

20. If you were soldering MOS solid-state devices you would not use:

   a. a soldering iron   **b.** rosin-core solder

   c. flux   **d.** a soldering gun

ANSWERS TO FILL-IN QUESTIONS AND SELF-CHECKING QUIZ

Experiment 1:   **(1)** cut, nick   **(2)** oxides   **(3)** tinned   **(4)** insulation

Experiment 2:   **(1)** mechanical, electrical   **(2)** close   **(3)** smaller

Experiment 3:   **(1)** tinned   **(2)** sponge   **(3)** solvent, alcohol   **(4)** clean
(5) too little solder, too much solder, fractured joint

Experiment 4:   **(1)** cut off   **(2)** flux   **(3)** capillary   **(4)** vacuum   **(5)** pulse

Self-Checking Quiz: **(1)** b   **(2)** d   **(3)** d   **(4)** b   **(5)** b   **(6)** b   **(7)** c   **(8)** a
(9) c   **(10)** b   **(11)** b   **(12)** a   **(13)** a   **(14)** c   **(15)** d
(16) c   **(17)** c   **(18)** b   **(19)** b   **(20)** d

Unit 7

Printed Circuit Board Assembly

INTRODUCTION

A *printed circuit board* (PC board) is a flat electrically insulated board with thin copper circuit patterns that connect together the components placed on it. Modern electronic components are very small and sensitive to physical stress and vibration. Printed circuit boards provide a sturdy base for mounting these small components and allow many components to be mounted in a small area. This *high packing density*, where more components are placed within a given area, is a particular advantage of producing smaller and lighter electronic equipment. The copper circuit patterns reduce standard wiring considerably, and in many cases are the only possible way to connect very small components. Printed circuit boards are largely responsible for the production of smaller and more reliable electronic equipment.

UNIT OBJECTIVES

Upon completion of this unit, you will be able to:

1. Describe how a printed circuit board is produced.
2. List the various types of printed circuit boards.
3. Mount axial and radial components on a PC board.
4. Mount multilead components and ICs on a PC board.
5. Perform the soldering techniques used for PC boards.
6. Explain surface mount technology components.
7. Describe the reflow soldering technique used in surface mount technology.
8. Define terms used in printed circuit board fabrication.

7-1a PRINTED CIRCUIT BOARD CONSTRUCTION

As shown in Figure 7–1, a basic printed circuit board has the copper circuit pattern or foil on one side of the board. Holes are drilled through pads or terminals in the foil and board. The component leads are pushed through the holes from the other side of the board. The leads are then soldered to the copper foil to complete the circuit connections.

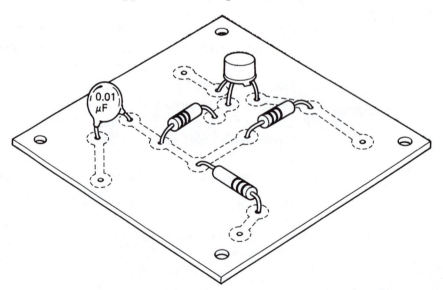

Figure 7–1   Basic printed circuit board.

The foil pattern takes various shapes, depending on the function of the circuit. Figure 7–2 shows an example of the shapes of circuit elements. Heat sinks are used to dissipate heat from components and

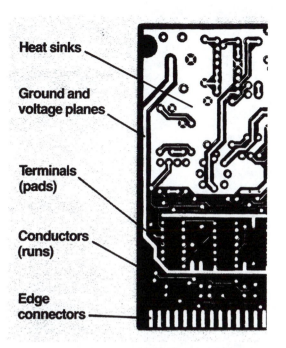

Figure 7–2 Shapes of circuit elements. (Permission to reprint granted by PACE Incorporated, 9893 Brewers Court, Laurel, MD, 20707.)

require large areas. Voltage and ground lines or planes to which components are attached are long and slender in length and will follow a path so as to provide power to all components on the board. Terminals or pads are the points that are drilled with holes to accommodate the component leads. Conductors or runs are thin foil strips between components. Edge connectors are part of the foil that connects the circuits of the board to a special plug, which in turn connects to other boards to form a system.

7–1a   Base or Substrate Material

The base material of a printed circuit board, referred to as the *substrate* or *laminate,* is made of phenolic paper, epoxy paper, and epoxy glass. The metal foil, referred to as the *cladding,* is usually made of copper, but other types of metal may be used. The substrate board is produced first and then the cladding is bonded to it.

7–1a.2   Method of Producing Circuitry

There are several methods for producing the conductive pattern on a printed circuit board. The circuit pattern connecting the components must be placed on the copper foil. The pattern may be drawn with a chemical, applied with special circuit tape, then produced by silk-screening or photographic means. Photographic methods are normally employed where many of the same types of board must be produced. Figure 7–3 illustrates a general method for producing a printed circuit board.

The copper side of the board must first be cleaned of the oxide layer with chemical cleaners and then rinsed in water (Figure 7–3a). A clear-to-amber-colored material that resembles and handles similar to varnish or lacquer is then painted or sprayed on top of the copper (Figure 7–3b). When exposed to light of the proper wavelength, this material, called *photoresist* or simply *resist*, is chemically changed in its solubility to certain solvents or developers. The resist must be applied in a dark area so as not to be exposed to ambient light. A preheat step normally follows in the process, to dry the resist (Figure 7–3c).

The next step is to photo-expose the circuit pattern onto the board

(Figure 7–3d). With a negative-acting resist the portions beneath the pattern that are not exposed to light become hardened and are not soluble in the developing solution. Positive-acting resist is also available, where the portion exposed to light will be soluble in solutions. A postheat step may be performed next but is not always required.

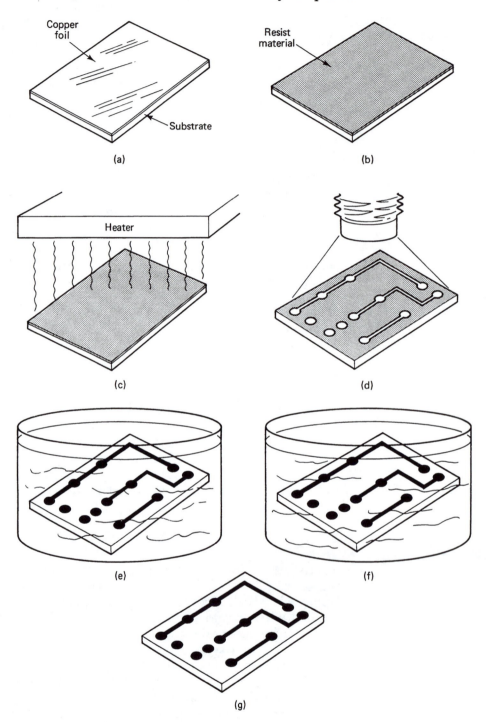

Figure 7–3 General method of printed circuit board construction: (a) clean copper foil; (b) resist application; (c) preheat; (d) photo-exposure; (e) etching; (f) cleaning; (g) finished PC board.

The exposed board is now placed into a solution called the *etchant*, which dissolves the copper and resist exposed to light (Figure 7–3e). The etchant solution used most often is ferric chloride ($FeCl_3$). This solution removes the exposed areas of the copper, leaving the substrate material underneath. The copper under the hardened resist remains, producing the circuit pattern. The board is then placed into a cleaning solution to remove any traces of resist that may interfere with soldering (Figure 7-3f).

The copper pattern surfaces are usually covered with a thin coating of solder (a tin–lead composition) to reduce the effects of oxidation and improve the soldering process of components. The finished printed circuit board is then dried and ready for assembly (Figure 7-3g).

7–1a.3 Through-Holes for Components

A printed circuit board has holes drilled in it which accept the various components for mounting. There are two types of hole configuration, shown in Figure 7–4. The *unsupported hole* is simply drilled through a pad and the substrate. The component is mounted from the substrate side. The *supported hole* has its wall lined or plated with copper so that it forms a continuous electrical path, called a *through-connection* or *through-hole,* running through the board from one side to the other. A plated through-hole is also known as a *reinforced hole.*

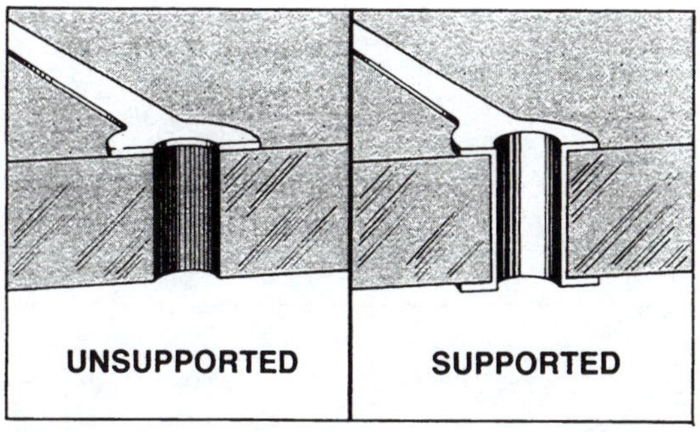

Figure 7–4 Typical hole configurations. (Permission to reprint granted by PACE Incorporated 9893 Brewers Court, Laurel, MD, 20707.)

UNSUPPORTED   SUPPORTED

In addition to plated through-holes, *eyelets* and *funnelets* may be used for added strength, as shown in Figure 7–5. Eyelets and funnelets are made of soft copper and are placed in printed circuit board holes for additional strength. One end of the eyelet or funnelet comes with a manufactured head or rim. The device is then placed into the hole and the other end is pressed outward with a special tool in a process called *swaging.*

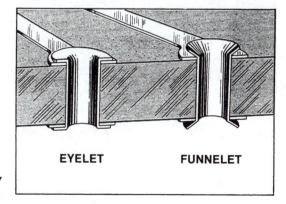

Figure 7–5 Eyelets and funnelets. (Permission to reprint granted by PACE Incorporated, 9893 Brewers Court, Laurel, MD, 20707.)

EYELET   FUNNELET

7–1a.4 Types of Printed Circuit Boards

Printed circuit boards are classified as to their type of construction. A *single-sided board,* the basic type of board used, has all the copper foil circuitry on one side, as shown in Figure 7–6. The holes through the pads are normally unsupported. The components are mounted on the nonconductive side of the board with their leads pushed through the holes. The leads are then soldered to the copper foil.

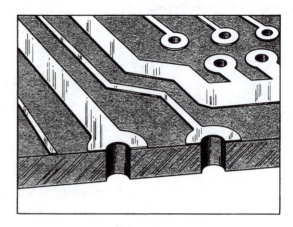

Figure 7–6 Single-sided board. (Permission to reprint granted by PACE Incorporated, 9893 Brewers Court, Laurel, MD, 20707.)

A *double-sided board* has circuitry on both sides of the board. In most applications there is a continuous conductive path between the circuitry on both sides, as shown in Figure 7–7a. This path is provided via a through-hole connection. The leads of the components are soldered from the circuit side of the board, but solder flows through the hole, providing a continuous solder joint.

A less used double-sided board with unsupported holes is shown in Figure 7–7b. There is no plated through-hole or other mechanical feed-through; the component lead performs this function. The lead is soldered on both sides of the board to assure a good connection. This type of board is more susceptible to solder joint cracking due to thermal stresses and is difficult to repair because each side must be desoldered independently.

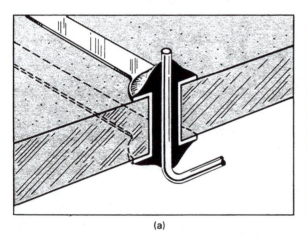

(a)

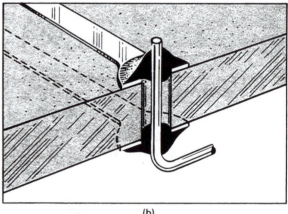

(b)

Figure 7–7 Double-sided board: (a) supported hole; (b) unsupported hole. (Permission to reprint granted by PACE Incorporated, 9893 Brewers Court, Laurel, MD, 20707.)

A multilayer board is similar in construction to a double-sided board but has other internal conductor planes sandwiched in layers within the board. These layers are connected together in many locations of the board via through-connections as shown in Figure 7–8. This type of board is more difficult to solder and repair and requires additional time to complete a solder operation.

Although not as common as the rigid substrate type of printed circuit board, flexible wiring assemblies are being used in more specialized equipment. A *flexible circuit* consists of etched conductors of electrolytic or rolled copper laminated to one or more layers of a flexible insulating base of plastic material. Very small components are normally used with flexible circuits.

A single-sided PC board is shown in Figure 7–9. The component side

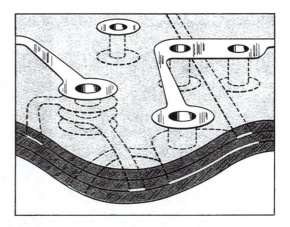

Figure 7–8 Multilayer board. (Permission to reprint granted by PACE Incorporated, 9893 Brewers Court, Laurel, MD, 20707.)

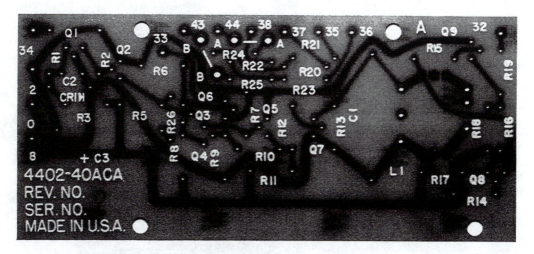

Figure 7–9 Single-sided PC board with component identification.

of the board is facing you. Component placement is indicated by letters and numbers. The letter indications preceding the numbers identify the following:

R stands for resistors
C stands for capacitors
CR stands for rectifier

L stands for inductor
Q stands for transistor

The other letters and numbers represent test or reference points for technicians to use in troubleshooting. Notice the dark shadow area lines, which are the circuit patterns on the reverse side of the board.

With a double-sided PC board, the component side also has some circuit patterns, as shown in Figure 7–10a. The placement of the components is indicated and you can see the dark circuit patterns on the reverse side. Figure 7–10b shows the same board with the components mounted.

7–1b MOUNTING PASSIVE COMPONENTS

Mounting axial-lead components such as resistors and capacitors into a PC board requires specific procedures. The leads of the components must be cleaned of oxidation before soldering begins. Fine abrasive materials such as sandpaper or rubber sticks can be used, but care must be taken not to clean the leads so much as to wear away the metal on the leads. A rubber stick is probably the best to use and can also be applied to the

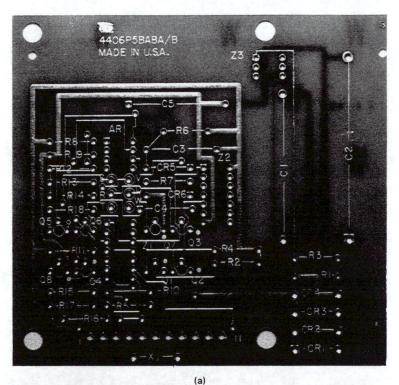

(a)

(b)

Figure 7–10 Double-sided PC board: (a) shadow lines are circuit pattern on reverse side; (b) with components mounted.

pads on a PC board. Gold-plated leads are very soft and should be cleaned with solvent and tissue only to prevent wearing away the plating.

7–1b.1 Forming Component Axial Leads

Forming the leads to fit into the required holes on the PC board is a very important task. Once the distance between the holes is established, the component leads can be formed in several ways. One way to form the leads is with a pair of long nose pliers, as shown in Figure 4–18. However, the leads cannot be scratched or scarred by the plier jaws. If the jaws are serrated, they must be covered with heat-shrinkable tubing or other suitable material so as not to damage the leads. Other lead-forming tools are designed for various sizes. The component is placed at the desired point of the tool and the leads are bent across the tool. One type of component lead-forming tool is shown in Figure 7–11.

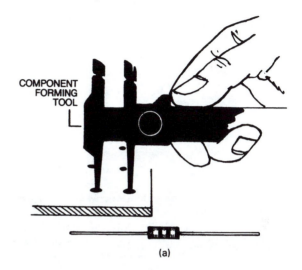

COMPONENT
FORMING
TOOL

(a)

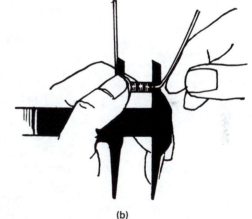

Figure 7–11 Component lead-forming tool: (a) measuring distance between holes; (b) forming leads on tool. (Permission to reprint granted by PACE Incorporated, 9893 Brewers Court, Laurel, MD, 20707.)

(b)

The two tapered measuring probes of the tool are first placed into the holes where the component is to be mounted. The distance between the holes is then locked into the tool with a knurled screw. The component is placed into slots across the opposite end of the tool and the leads are formed.

7–1b.2 Methods of Lead Termination to a PC Board

The leads of all types of components can be mounted to a PC board in several ways, as shown in Figure 7–12. The *straight-through method* is the easiest to use because there are no bends in the leads after they pass through the holes. This type of lead is also easy to remove.

With the *clinched method,* the lead is left long enough to be bent 90° onto the pad or conductor. This helps to stabilize the component on the board so that it will not move during the soldering process. This type of termination is more difficult to remove. The *semiclinched lead* has about a 45° bend in the lead, which makes it easier to straighten out and remove.

STRAIGHT THROUGH

CLINCHED

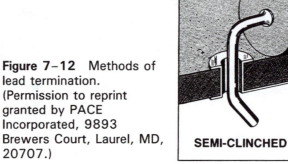

Figure 7–12 Methods of lead termination. (Permission to reprint granted by PACE Incorporated, 9893 Brewers Court, Laurel, MD, 20707.)

SEMI-CLINCHED

SPADED

A *swaged* or *spaded lead* is a lead that extends beyond the hole and is then flattened so that it is wider than the diameter of the hole. The swaged end helps to hold the component in the proper position during handling and soldering. However, removing this type of lead requires more time and special procedures.

7–1b.3 Mounting the Component

After the leads of the component are formed, the component is placed into the PC board as shown in Figure 7–13a. The component is held in place and the leads protruding through the holes are trimmed (cut) to the proper length, depending on the type of termination used (Figure 7–13b). If the termination is the clinched type, the end of the lead is bent over the conductor with a nonmetallic tool so as not to scratch the lead (Figure 7–13c).

Figure 7–13 Mounting axial-lead components: (a) inserting component; (b) trimming for clinched lead; (c) clinching with nonmetallic tool. (Permission to reprint granted by PACE Incorporated, 9893 Brewers Court, Laurel, MD, 20707.)

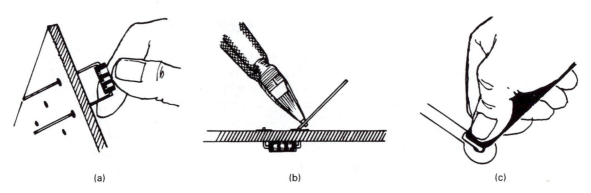

(a)   (b)   (c)

7–1b.4 Solder Procedures for Component Leads in PC Boards

Soldering components onto PC boards requires considerable caution, due to the heat sensitivity of both the board and components. The copper foil, which is about 0.0014 to 0.0028 in. thick, is sensitive to heat. A hot soldering iron tip can rapidly deform or lift the board's pads or conductors if too much pressure is applied. Also, if the iron tip is left on one spot for too long, the conductor may lift off the board. The soldering iron should never be pushed down on the pad, but merely rests on it.

Figure 7–14 shows the general procedure for soldering leads on a PC board. The solder and iron are placed on the same side of the lead initially, to create a solder bridge. The solder is then moved to the opposite side of the lead and allowed to flow for a short time. The solder is removed and the iron is removed with a wiping action. For a clinched lead the wipe is in the direction of the lead, and for a straight-though lead the iron is wiped in an upward motion. The solder joint is then cleaned with a solvent such as alcohol to remove any flux residue. Finally, the solder joint is inspected for acceptable quality.

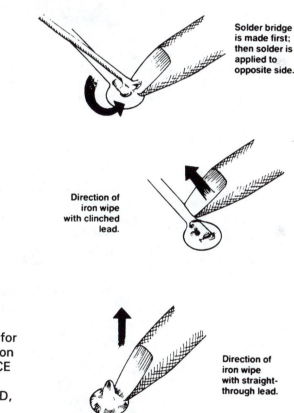

Solder bridge is made first; then solder is applied to opposite side.

Direction of iron wipe with clinched lead.

Direction of iron wipe with straight-through lead.

Figure 7–14 Procedure for soldering leads. (Permission to reprint granted by PACE Incorporated, 9893 Brewers Court, Laurel, MD, 20707.)

Figure 7–15 shows some examples of acceptable straight-through lead solder joints. Notice that the surface of the solder is smooth and well feathered out to the edge. There should be no evidence of pits or holes and residue flux.

Acceptable clinched lead solder joints are shown in Figure 7–16. Notice the slight concave appearance of the solder and that the shape of the wire can still be seen beneath the solder. The length of solder flow along the lead should not be excessive. A long solder flow indicates too long a heat dwell time and the possibility of heat damage at the joint

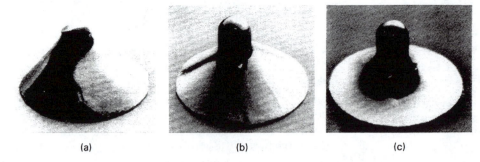

(a)   (b)   (c)

Figure 7–15 Axial-lead straight-through joints: (a) minimum acceptable solder; (b) preferred joint; (c) maximum acceptable solder. (Permission to reprint granted by PACE Incorporated, 9893 Brewers Court, Laurel, MD, 20707.)

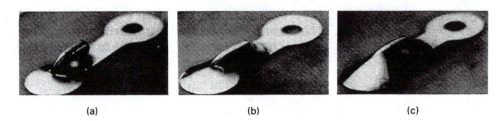

(a)   (b)   (c)

Figure 7–16 Axial-lead clinched joints: (a) minimum acceptable solder; (b) preferred joint; (c) maximum acceptable solder. (Permission to reprint granted by PACE Incorporated, 9893 Brewers Court, Laurel, MD, 20707.)

or component. You should try to make solder joints look like the preferred joint shown in Figure 7–16b.

Unacceptable solder joints are shown in Figure 7–17. A *rosin joint* (Figure 7–17a), indicates that too little heat was applied to the joint and there is still a quantity of solidified flux between the wire and the terminal. Sometimes the flux may appear on the solder surface itself.

A *cold solder joint* (Figure 7–17b), which will appear not to cover the joint smoothly, is a result of withdrawing the heat of the iron too

Figure 7–17 Unacceptable axial-lead joints: (a) rosin joint; (b) cold joints; (c) disturbed joint; (d) overheated joint. (Permission to reprint granted by PACE Incorporated, 9893 Brewers Court, Laurel, MD, 20707.)

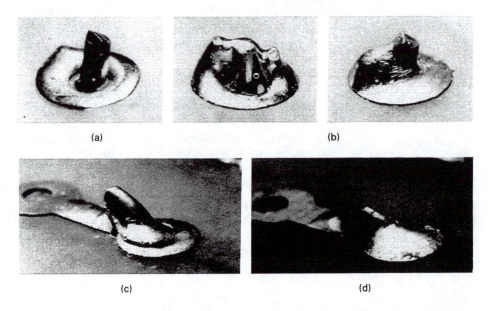

(a)   (b)

(c)   (d)

soon. The solder has not had a chance to become fully liquid and flow freely over the entire joint.

A *disturbed joint* (Figure 7–17c) will appear frosty and granulated. It may also show signs of cracks, indicating movement of the lead or wire during solder solidification.

An *overheated joint* (Figure 7–17d), which is chalky, dull, or crystalline in appearance, may be pitted on the surface. This type of joint results from repeated attempts at soldering. A joint such as this, that will not wet properly, is usually due to contamination or lack of sufficient flux.

After the solder joint has been thoroughly cleaned, it must be inspected. The best method of inspecting a joint is to rotate it under an overhead light. This movement should reveal any pits, discontinuities, and/or discolorations on the surface of the solder.

7–1b.5 Mounting Radial-Lead Components

Radial-lead components are mounted upright, as shown in Figure 7–18a. The component may be mounted flush with the board or may be raised a small amount above the board. Axial-lead components can also be mounted in an upright position, as shown in Figure 7–18b. The long lead is formed around the component before it is inserted into the board. Components mounted in this manner require less space on a PC board and allow a higher packing density.

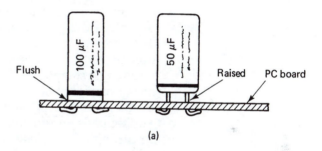

(a)

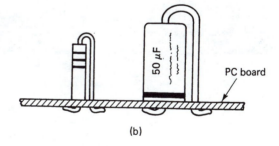

Figure 7–18 Mounting radial-lead components: (a) radial-lead component mounted upright; (b) axial-lead component mounted upright.

(b)

Summary of Axial-Lead Component Assembly Procedures

1. Clean the component leads.
2. Clean the pads on the PC board.
3. Measure the holes where the component is to be inserted.
4. Form the leads of the component.
5. Insert the component into the PC board.
6. Trim the leads for required termination.
7. Clinch the component leads, if required.
8. Solder the component leads to the PC board.

9. Clean the solder joint with solvent.

10. Inspect the solder joint.

7–1c MOUNTING ACTIVE COMPONENTS

Mounting and soldering active components such as diodes, bipolar transistors, FETs, LEDs, ICs, and related devices require extreme care and patience. These devices are very sensitive to heat and can be destroyed very easily. The soldering iron tip should be on the joint and lead no longer than 1 to 2 seconds. In some cases heat sinks can be placed on the leads during soldering, which restricts the heat from entering the main body or package of the device.

7–1c.1 Mounting Multilead Components

Regular diodes are axial two-lead devices and are mounted on PC boards similar to resistors and capacitors. The LED is a radial lead component and is not difficult to mount on a PC board. Some of the methods used for mounting LEDs were described in Section 5–1c.11 and shown in Figure 5–16. Rectifier modules containing diodes have four radial leads and mount upright. The main consideration with mounting a rectifier is to observe polarity and place the positive (+) and negative(−) leads into the proper terminals in the PC board.

Transistors and FETs usually have three radial leads and mount on a PC board as shown in Figure 7–19. The leads go directly through the holes in the board and are soldered to the pads on the circuit side. This component may be mounted flush with the board (Figure 7–19a). Another method uses a Transipad under the component body (Figure 7–19b). The component leads go through the Transipad and then through the appropriate holes in the board. The Transipad acts as a distance gage as well as providing some heat sinking. Some components are soldered to offset pads on the circuit side of the board (Figure 7–19c). After passing through the holes the leads are bent to lie flat on the corresponding pads.

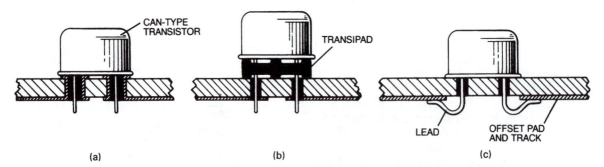

Figure 7–19 Mounting transistors: (a) through-plate mounting in plated-through hole (third lead is not shown); (b) use of Transipad for spacing; (c) leads formed to reach offset pads. (Permission to reprint granted by PACE Incorporated, 9893 Brewers Court, Laurel, MD, 20707.)

With multilead components such as ICs with TO–5 packages, more time is required to align the leads properly before they are placed into the board. Figure 7–20 shows one method for preparing a multilead package. The leads are first flared with the end of a rubber abrasive stick (Figure 7–20a). Next, a pair of diagonal cutters are used to cut the leads as the component is rotated (Figure 7–20b). Cut very little off the first lead, because each following lead is trimmed shorter than the preceding lead. This technique makes it easier to mount the component into the

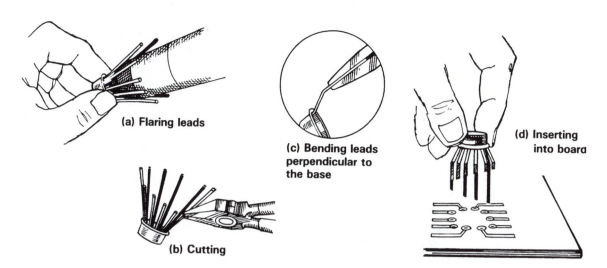

(a) Flaring leads

(c) Bending leads perpendicular to the base

(d) Inserting into board

(b) Cutting

Figure 7-20 Mounting multilead components. (Permission to reprint granted by PACE Incorporated, 9893 Brewers Court, Laurel, MD, 20707.)

holes in the PC board. Next, the leads are bent perpendicular to the base (Figure 7-20c). Finally, the component is aligned and the longest lead is placed into the board first. Then rotate the component slightly as you insert each lead in turn (Figure 7-20d). The board is now turned over and each lead cut to the proper length and soldered to the circuit. Figure 7-21 shows the preferred solder joints for transistors and other multilead components. Notice the smooth shiny surface of the solder.

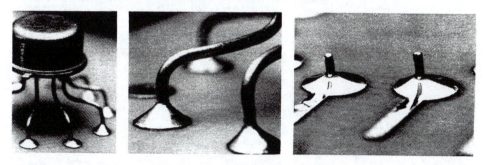

Figure 7-21 Preferred joints for transistors and other multilead components. (Permission to reprint granted by PACE Incorporated, 9893 Brewers Court, Laurel, MD, 20707.)

7-1c.2 Mounting DIP Integrated Circuits

A major portion of electronic assembly involves the mounting and soldering of DIP ICs. These ICs are mounted onto double-sided boards having plated-through holes. The main consideration here is to ensure that sufficient solder is applied to fill the hole and wet the lead surfaces on both sides of the board. Figure 7-22 illustrates the steps in mounting and soldering a DIP IC.

The DIP comes with its leads slightly sprung outward; therefore, it is necessary to bend the leads so that they can be inserted easily in the holes. Bending can be done by gently holding the body of the package with a pair of long nose pliers and then pressing both rows of leads sideways down onto a flat surface so that the leads are more perpendicular to the body of the package. The DIP IC is then placed into the board and held in place with one finger as the board is turned over. Next, two leads on the opposite corners are clinched outward to keep the DIP from moving during soldering. Special tools can be used to straighten and align the leads (Figure 7–22a). In this case the tool holds the leads compressed as the DIP is placed into the board. When the tool is released the leads spring outward to the side walls of the holes and hold the DIP firmly in place while the board is turned over for soldering.

Since the leads on a DIP IC are small, a small soldering iron tip is used. The iron tip is applied to one side of the lead and pad and the solder is placed on the other side (Figure 7–22b). A pool of solder builds up on the top side of the pad first, so the iron tip must be kept on the joint until the pool drops through the hole to the bottom side. After the solder flow-through, a little more solder is added to the top side to mount it up again (Figure 7–22c). However, the DIP IC is very sensitive to heat and the iron should not be left on the joint longer than 3 seconds. Alternate leads along the row should be soldered so as to prevent heat buildup and damage to the body of the IC. The procedure is the same for soldering DIPs onto multilayer boards.

Acceptable solder joints for DIP ICs are shown in Figure 7–23. Notice how the solder has flowed upward through the holes onto the leads of the DIP. There are no air pockets, flux pockets, or voids where contaminants could gather.

Important Points to Remember When Soldering PC Boards

1. Make sure that the leads and pads are clean.
2. Use sufficient flux at the joint.

Figure 7–22 Soldering DIP ICs. (Permission to reprint granted by PACE Incorporated, 9893 Brewers Court, Laurel, MD, 20707.)

a. Straightening leads.

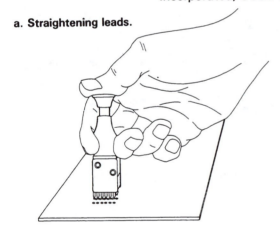

b. Solder builds up on only one side at first, so iron must be kept on until the pool drops, indicating the solder has flowed through the hole.

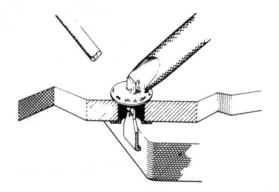

c. After flow-through, a little more solder is added to mount it up again.

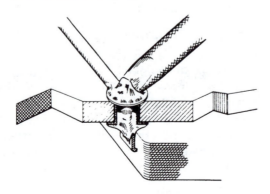

3. Do not push the soldering iron tip on the foil; merely rest it there.
4. Keep the heat dwell time as short as possible (usually less than 3 seconds) to obtain an acceptable solder joint.

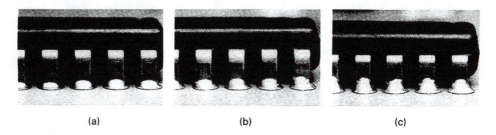

(a)   (b)   (c)

Figure 7–23 DIP IC solder joints: (a) minimum acceptable solder; (b) preferred joints; (c) maximum acceptable solder. (Permission to reprint granted by PACE Incorporated, 9893 Brewers Court, Laurel, MD, 20707.)

7–1d SURFACE MOUNT TECHNOLOGY

Consumers constantly demand more features from their electronic products and desire the equipment to be smaller and easier to handle. In an attempt to meet this demand, manufacturers must reduce the size of electronic circuits more and more.

As integrated circuits become more complex, the size of their packages requires more connections to the outside world—hence the package becomes larger. In reality, the integrated circuit or chip that it is fabricated on increases very little. It is the package with the increase in leads that becomes larger. The older standard DIP IC had leads that are 0.100 in. apart on center. Even providing leads or pins on all four sides is insufficient for some circuits. On newer IC packages the leads have been reduced to 0.050 in. apart on center, which allows a 30 to 60% reduction in the size of the package. This means that more ICs can fit in the same amount of PC board space.

Standard DIP ICs belong to the *insertion mount technology* (IMT) category, because they are inserted into holes that extend through the PC board. This means that circuitry is usually limited to one side of the board or must be routed around other IC packages. In a newer technique called *surface mount technology* (SMT) all the leads of the package are connected on one side of the PC board without going through holes. This allows separate circuits to be constructed on each side of the board. In effect, you can have the circuitry of two boards on one board. Wire jumpers can run from one side to the other side of the board if required to connect the circuits. Also, since the leads of a SMT device do not penetrate the PC board, boards are both cheaper to manufacture and more reliable in operation.

7–1d.1 Types of Surface-Mountable Components

Components used with surface mount technology may be referred to as *surface mount devices* (SMDs) or *surface mount components* (SMCs). There are also more specific names given to individual devices.

7–1d.1a Flat-Pack IC

One of the oldest SMDs is the flat pack, shown in Figure 7–24. The *flat pack*, an IC with very thin copper- or gold-plated leads, is usually *planar-mounted*, which means that all connections are on the same side of the

Figure 7–24 Flat pack.

Figure 7–25 Reflow technique for soldering flat packs. (Permission to reprint granted by PACE Incorporated, 9893 Brewers Court, Laurel, MD, 20707.)

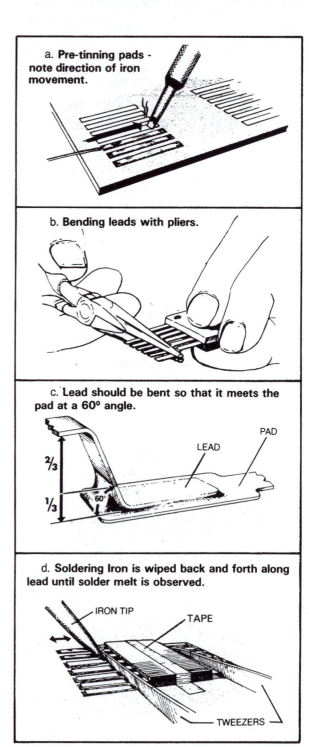

a. **Pre-tinning pads - note direction of iron movement.**

b. **Bending leads with pliers.**

c. **Lead should be bent so that it meets the pad at a 60° angle.**

PAD

LEAD

2/3

1/3

60°

d. **Soldering Iron is wiped back and forth along lead until solder melt is observed.**

IRON TIP

TAPE

TWEEZERS

PC board. These leads are very delicate and the flat pack must be handled and stored with extreme care.

Since the leads of the flat pack and other SMCs are small in size and delicate in nature, normal soldering procedures cannot be used. Most SMT devices are mounted using the *reflow soldering technique*. With this technique, the pads on the board are pretinned, as sometimes are the leads of the device to be mounted. A step-by-step procedure for mounting flat packs is shown in Figure 7–25.

1. *Figure 7–25a:* The pads of the PC board are cleaned and then pretinned. A small soldering iron tip and a small solid-core area of solder, about 0.010 in. in diameter, are used. Notice how the iron and solder are applied to the pads and the direction of movement of the iron tip. After all the pads are pretinned, the area is cleaned with a solvent to remove used flux.

2. *Figure 7–25b:* Some flat packs come with their leads straight out and must be formed before being mounted to the PC board. A pair of long nose pliers can be used to form the leads.

3. *Figure 7–15c:* Each lead requires two bends and should contact the pad at a 60° angle. The leads are cut so that they fit lengthwise along the pad, but not to the end. Also, the leads must not be too short. Figure 7–26a shows an acceptable lead length for a flat pack. After forming and cutting, the leads are cleaned with a solvent. In some cases the leads may also be pretinned.

4. *Figure 7–25d:* Before soldering, new flux is placed on the leads. The reflow soldering technique involves placing the flat pack on top of the pads where it is to be mounted. A thin piece of masking tape can be used to hold the component in place temporarily while soldering. After positioning the flat pack the leads at opposite ends are reflow-soldered and then the remaining leads are reflow-soldered alternately to prevent heat buildup. To perform the reflow-solder technique, a small tool or pair of tweezers is used to apply a downward pressure on the upper lead surface as the soldering iron tip is moved back and forth along the lead. The pretinned solder on the pad will melt and flow up onto the lead. When this action is observed, the irons should be removed, using a wiping motion toward the end of the lead. The wiping motion prevents any lifting or movement of the lead. After all the leads are soldered, the remaining flux is removed with solvent and the joints are inspected. A proper joint will have solder covering the heel of the lead, but only about one-third up the lead and never beyond the upper bend.

Solder joints for flat packs are shown in Figure 7–26. Acceptable leads must be parallel to the pad and centered on it. The joint is completely wetted and forms a secure bond. Joints are unacceptable if the leads are not properly centered and if they overhang the edge of the pad.

A "lap flow" tool is specially designed for reflow soldering applications. A foot switch controls the time the iron is on. The tool is placed on the joint and holds the lead clamped in the proper position on the pad. The switch is then activated and heat is provided to the joint. Once the wetting action has occurred, the heat is switched off, but the tool remains on the joint, holding it in place until the solder solidifies.

7–1d.1b Small Outline Transistor

A version of the discrete transistor is available in SMC form. The *small outline transistor* (SOT) is shown in Figure 7–27. The standard configura-

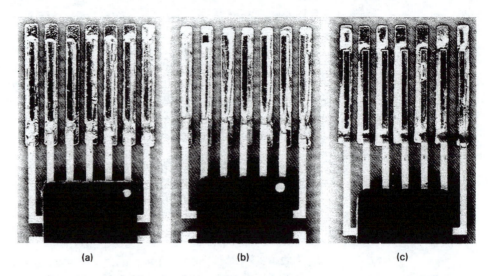

(a)   (b)   (c)

Figure 7–26 Flat-pack solder joints: (a) preferred joints; (b) maximum acceptable lead overhang; (c) unacceptable lead overhang. (Permission to reprint granted by PACE Incorporated, 9893 Brewers Court, Laurel, MD, 20707.)

tion has two leads on one side and one lead on the other side (Figure 7–27a). A side view (Figure 7–27b) shows that the SOT has leads shaped like gull wings. The SOT operates in the same manner as the conventional package transistor, but it is much smaller. The SOT–23 is about $\frac{1}{10}$ in. on a side. In actual mounting on a PC board, it occupies about one-fifth of the board space required by a conventional TO–92 transistor package. Other SOT package configurations are shown in Figure 7–27c, d, and e.

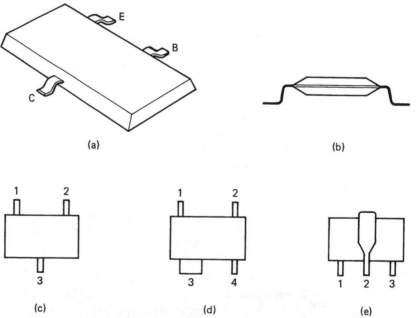

Figure 7–27 SOT packages: (a) typical SO transistor package; (b) end view showing gull-wing leads; (c) SOT 23; (d) SOT 143; (e) SOT 89.

7–1d.1c Small Outline Integrated Circuit

The integrated circuit can also be found in SMD form and is called a *small outline integrated circuit* (SOIC). The package has the gull-wing leads and is available in DIP form or with leads on all four sides, as shown in Figure 7–28.

Figure 7–28 Small outline integrated circuit (SOIC).

7–1d.1d Plastic-Leaded Chip Carrier

Another type of SMD configuration is the *plastic-leaded chip carrier* (PLCC), shown in Figure 7–29a. The leads of a PLCC package are similar to that of a SOIC package, except that they are rolled up under the molded plastic body in a J shape, as shown in Figure 7–29b. With the leads beneath the package, the size of the IC is reduced and greater component density on a PC board is made possible. PLCCs are available in 18-, 20-, 28-, 44-, 52-, 68-, and 84-pin packages. Another advantage of the PLCC is that it can be placed into a socket, whereas other types of SMDs cannot.

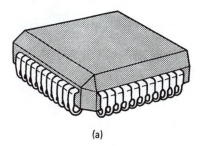

(a)

Figure 7–29 Plastic-leaded chip carrier (PLCC): (a) typical PLCC with leads on all four sides; (b) cross-sectional view of J-lead.

(b)

7–1d.1e Leadless Ceramic Chip Carrier

The *leadless ceramic chip carrier* (LCCC) is the most rugged of the SMDs. Its main feature is that it has no leads; rather, the metallic contacts are molded into its ceramic body as shown in Figure 7–30. The leads may be in DIP form or on all four sides.

PLCC and LCCC devices are very difficult, if not impossible, to solder with conventional soldering tools. These types of components require

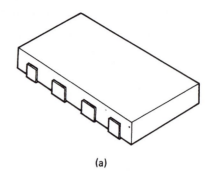

(a)

Figure 7-30 Leadless ceramic chip carrier (LCCC): (a) typical LCCC packages; (b) cross-sectional view showing external connections.

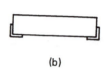

(b)

specialized soldering tools and machines for proper mounting to a PC board. These tools and special soldering techniques are discussed in subsequent units.

7–1d.1f SMT Chip Components

SMT passive electronic components are also manufactured and referred to as *chip components* because of their very small size. A *chip resistor* is a SMD and can range in size from 0.08 to 0.30 in. in length, as shown in Figure 7–31. The resistive material is deposited on a ceramic substrate, as with thick-film resistors. A glass epoxy is placed over the resistive material to protect it. The chip resistor is leadless but has external electrodes for attaching it to the circuit. The internal electrode attaches to the resistive material. The secondary electrode is made of nickel and protects the resistive material from the heat during the soldering process. The external electrode has a coating of solder and is easily soldered to the PC board. The resistive range of chip resistors is from 10 Ω to over 10 MΩ.

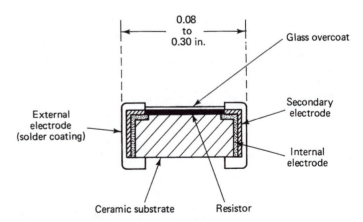

Figure 7-31 Chip resistor.

The chip capacitor resembles the chip resistor in physical appearance, but has a different type of construction, as shown in Figure 7–32. The ceramic chip capacitor is the most used and consists of a multilayer of metal film as the plates and the ceramic as the dielectric. The end terminals are similar to those of the chip resistor, with an internal electrode, a nickel secondary electrode, and the external solder-plated electrode. Capacitance values range from 1 pF to 1μF. Other types

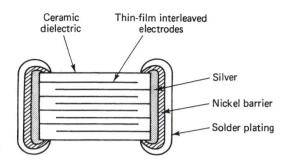

Figure 7–32 Chip capacitor construction.

Ceramic dielectric — Thin-film interleaved electrodes — Silver — Nickel barrier — Solder plating

of capacitors are also available. Aluminum electrolytics range from 1.5 to 4.7 μF and tantalum dielectrics range from 0.1 to 100 μF. Figure 7–33 shows some types of SMC packages. The beveled edge or pointed end of the polarized component usually indicates the positive (+) terminal.

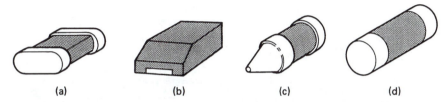

(a)   (b)   (c)   (d)

Figure 7–33 Types of SMT chip components: (a) standard chip; (b) polarized chip; (c) polarized cylindrical; (d) standard cylindrical.

Chip components also include inductors, switches, relays, crystals, and crystal filters. Remember, most of these devices are less than $\frac{1}{4}$ in. in length. Figure 7–34 and 7–35 shows the relative size of SMDs.

Figure 7–34 SMD resistors and capacitors. Notice the SMD's size compared with a normal paper clip.

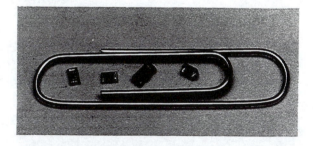

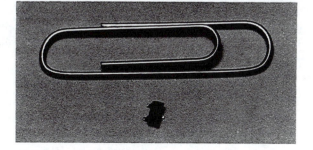

Figure 7–35 SMD bipolar transistor. Notice the SMD's size compared with a normal paper clip.

7–1d.1g   Mounting SMT Chip Components

Many SMT chip components can be mounted to a PC board using a very fine pointed soldering iron tip and the reflow solder technique. Figure 7–36 shows general procedures for mounting these components.

1. *Figure 7–36a:* The pads on the PC board are cleaned and pretinned.
2. *Figure 7–36b:* The terminals of the SMC are cleaned and the com-

ponent is placed on the pads. A small piece of masking tape may be used to hold the component in place.

3. *Figure 7–36c:* A pair of tweezers or a small tool is used to apply a downward pressure on the component while the soldering iron is applied to one end terminal. Once the solder liquefies and the wetting action is observed—where the solder runs up onto the terminal of the component—the iron is removed. The downward pressure is still applied by the tool until the joint has solidified. The tool is removed and the other terminal end of the component can now be soldered.

4. *Figure 7–36d:* The solder joints are cleaned with solvent and inspected. The solder joints should have a smooth, shiny-looking finish extending from the pad up along the component terminal end. You must be careful not to allow flux to build up and form between the solder and the terminal end.

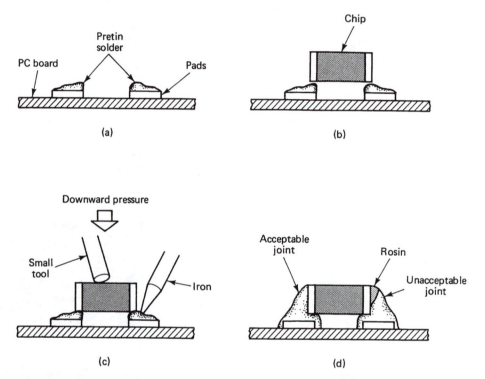

Figure 7–36 Soldering procedures for SMT chip components: (a) pretin pads; (b) place chip on soldered pads; (c) apply pressure to chip while soldering; (d) acceptable and unacceptable joints.

7–1d.1h   SMT Solder Flux and Conductive Adhesives

Reflow solder flux used with mounting SMCs consists of microscopic particles of solder suspended in the flux. This type of flux may be referred to as *solder paste* or *cream.* Small dots of solder cream are placed on the pads of the PC board. The SMCs are then placed on the pads and are held in place by the adhesive action of the cream. One advantage of using solder paste is the the placement of the SMCs is less critical, because when the solder melts, its surface tension tends to pull the SMCs back into position precisely over the pads.

Electrically conductive adhesives consist of a conductive powder suspended in a bonding base material, which can be used to mount SMCs to PC boards without solder. The powders may be of gold, silver, copper, nickel, carbon, and graphite. These conductive adhesives fall into two main classes:

1. *Thermosetting adhesives,* where the resulting bond is permanent and inflexible. The bonding materials used in this adhesive are epoxies, acrylics, and polyesters. The bond is cured or made permanent by a chemical catalyst, heat, or ultraviolet radiation. Repair or rework can be accomplished on thermosetting adhesives only after the bond is shattered or dissolved with a solvent.

2. *Thermoplastic adhesives,* where the bond can be flexible or reworked by the application of heat. The bonding material for this adhesive consists of nylon, polyimide siloxane, and a mixture of polymers. A thermoplastic adhesive bond is weaker than a thermosetting adhesive bond, but it is easier to remove and replace components simply with the use of heat.

In most cases, conductive adhesives are placed on the pads of the PC board and then the component is pressed onto the pads. The adhesive may contain the catalyst, and time is a factor in bonding the component properly to the board. In other cases, the catalyst is added to bond the component to the board.

There also exists *conductive inks,* which are used in some applications as circuit conductors for connecting components. Conductive inks are generally used in making prototype (first time) circuits, single circuits, or by hobbyists.

7–1e SOLDERING COMPONENT SOCKETS

Most IMT active components can be placed into accompanying sockets. Transistor sockets must be mounted in a specific direction so that the leads can be properly inserted, but DIP sockets usually can be mounted in either direction. Remember, PLCCs can be placed into special sockets. Refer to Figure 3–22, which shows various component sockets.

Soldering sockets involves the same technique as soldering terminal connections. The socket is pushed through the holes in the PC board. The board is turned upright with the solder connections of the socket protruding through the holes on the circuit side of the board. The soldering iron is placed on the joint and solder is then added. After wetting is accomplished, the solder is removed and the iron is taken off the joint with an upward wiping action. After all socket connections are soldered, the joints are cleaned with solvent and inspected. Sockets are not as sensitive to heat as are their respective components; however, care must still be used so as not to burn the PC board or surrounding components.

7–1f COMPONENT REPLACEMENT

Component replacement can be fairly easy or extremely difficult, depending on the complexity of the equipment and the types of PC boards used.

7–1f.1 Simple Component Replacement

An older method of component replacement, used only in an emergency situation, is shown in Figure 7–37. In the case of a PC board with ample room around the components, the old component can be removed by cutting it off at the leads while it is still mounted to the board. The remaining leads in the board are then formed into hooks. The new component

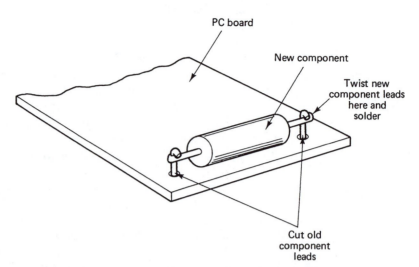

Figure 7–37 Simple
component replacement.

is then put in place and its leads wrapped around the hooks. Any excess
lead is cut off and the component is soldered into place.

Most component replacements within the electronics industry con-
form to the original standard of the circuit. Therefore, the procedures
described below are generally utilized and in many cases represent the
only methods available.

7–1f.2 Conformal Coating Removal

Many PC boards have a coating of epoxy, silicone, varnish, or some other
similar substance that completely covers all components and is referred
to as the *conformal coating*. This coating provides insulation to the components and board while protecting against mechanical shock and vibration, humidity, and fungus. The several types of conformal coatings can
be removed by various methods, such as solvent, heat (thermal parting),
abrasion with small grinding tools, and hot air. In many instances this
coating must be removed before the component can be replaced. After
the new component has been installed, the same type of conformal coating
must be applied to the area to return the PC board to its original condition to provide quality assurance.

7–1f.3 Removing Solder from Leads

Solder can be removed from components by using the wicking braid
technique described in Section 6–1b.10 and shown in Figure 6–19.
However, the best method to use on standard components is with a
vacuum desoldering tool. The size of the tool tip is very important for
performing an efficient job. The outside diameter (OD) of the tip should
not overextend the pad, as shown in Figure 7–38. The inside diameter
(ID) of the tip should be slightly larger than the component lead, to allow
molten solder to pass through when the vacuum draws the solder upward.

Figure 7–38 Desoldering
tip selection criteria.
(Permission to reprint
granted by PACE
Incorporated, 9893
Brewers Court, Laurel, MD,
20707.)

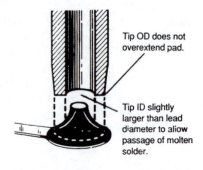

Tip OD does not
overextend pad.

Tip ID slightly
larger than lead
diameter to allow
passage of molten
solder.

Once the desoldering tool tip is placed on the component lead, movement is required to help loosen the solder, as shown in Figure 7–39. Round leads require a circular motion of the tip (Figure 7–39a) and flat leads, such as on a DIP IC, require a back-and-forth motion (Figure 7–39b).

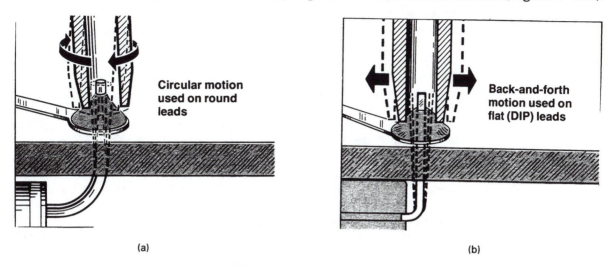

Circular motion used on round leads

Back-and-forth motion used on flat (DIP) leads

(a)

(b)

Figure 7–39 Desoldering component leads: (a) round lead; (b) flat lead (DIP). (Permission to reprint granted by PACE Incorporated, 9893 Brewers Court, Laurel, MD, 20707.)

In the case of desoldering a terminal, as shown in Figure 7–40, molten solder is drawn into the tool and then expelled several times to remove as much solder as possible from the terminal. A soldering iron is then used to heat the terminal as the wire is gently shaken off.

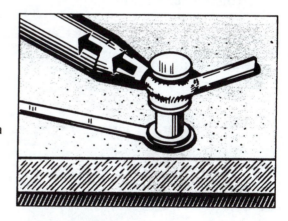

Figure 7–40 Desoldering a terminal. (Permission to reprint granted by PACE Incorporated, 9893 Brewers Court, Laurel, MD, 20707.)

7–1f.4 Replacing Components in a PC Board

Once the defective component is removed from the PC board, the area is then cleaned of excess solder and debris. If the new component has axial leads, the leads must be formed to fit the holes in the board (see Section 7–1b.1 and Figure 7–11). Methods of termination of the leads will depend on the original method used on the board (see Section 7–1b.2 and Figure 7–12). After the leads are formed, the component can be mounted in the PC board (review Sections 7–1b.3 and 7–1b.4 and Figures 7–13 to 7–17).

Radial lead components are usually easier to mount in a PC board since no forming of the leads is required. The leads are simply pushed through the holes and soldered (see Section 7–1b.5 and Figure 7–18). Replacing active components requires a little more effort because there are more than three leads on the components and special attention must be given to MOS devices. Be sure that all equipment is properly grounded, wear a ground wrist strap when working on MOS devices, and handle them with care (see Section 7–1c.1 and Figures 7–19 to 7–21).

Integrated circuits require considerable more removal time if they are *hard wired* (soldered to the board). Special desoldering tips are available which contact all of the pins at the same time. When the solder becomes molten, the IC is pulled out of the board. If a vacuum desoldering tool is used, each pin must have as much solder removed as possible. Then a soldering iron is placed on each pin beginning at one end, and the IC is gently rocked back and forth. This procedure is performed until all pins will move in the holes, and then the IC is removed from the board. The area of the board is cleaned and the new IC is put in place (see Section 7–1c.2 and Figures 7–21 to 7–23).

Surface mount devices are usually very difficult to work with. Extreme care and patience must be used when replacing these types of components. In many cases, special tools and equipment are used to remove and install SMDs. Remember to use the least amount of heat possible to do the job and to keep the dwell time to a minimum (see Section 7–1d.1a and Figures 7–25 and 7–26, and Section 7–1d.1g and Figure 7–36).

≡≡≡ **SECTION 7–2**
DEFINITION EXERCISES

Write a brief description of each of the following terms.

1. Printed circuit board _____

_____

2. High packing density _____

_____

3. Copper pattern or foil _____

_____

4. PC board heat sinks _____

_____

5. Ground and voltage planes _____

_____

6. Terminals or pads _____

_____

7. Conductors or runs _____

_____

8. Edge connectors _____

_____

9. Substrate _____

_____

10. Laminate _____

_____

11. Cladding _____

_____

12. Photoresist or resist _____

_____

13. Etchant _____

_____

14. Unsupported hole _____

_____

15. Supported hole _____

_____

16. Through-connection _____

_____

17. Through-hole _____

_____

18. Reinforced hole _____

_____

19. Eyelets and funnelets _____

_____

20. Swaging _____

_____

21. Single-sided board _____

_____

22. Double-sided board _____

_____

23. Multilayer board _____

_____

24. Flexible board _____

_____

25. Clinched _____

_____

26. Rosin joint _____

_____

27. Cold joint _____

_____

28. Disturbed joint _____

_____

29. Overheated joint _____

_____

30. Insertion mount technology (IMT) _____

_____

31. Surface mount technology (SMT) _____

_____

32. SMD _____

_____

33. SMC _____

_____

34. Reflow solder technique _____

_____

35. Flat pack _____

_____

36. Planar mounted _____

_____

37. SOT _____

_____

38. SOIC _____

_____

39. PLCC _____

_____

40. LCCC _____

_____

41. Chip components _____

_____

42. Gull-wing lead _____

_____

43. J-lead _____

_____

44. Lap flow tool _____

_____

45. Reflow solder flux _____

_____

46. Electrically conductive adhesives _____

_____

47. Thermosetting adhesive _____

_____

48. Thermoplastic adhesive _____

_____

49. Conductive inks _____

_____

50. Sockets _____

_____

≡≡≡ **SECTION 7–3**
   EXERCISES AND PROBLEMS

Complete this section before beginning the next section.

1. List the two basic parts of a PC board.

   a.

   b.

2. List at least six steps in the construction process of a PC board.

   a.

   b.

   c.

   d.

   e.

   f.

3. List four types of PC boards.

   a.

   b.

   c.

   d.

4. Name five of the shapes of circuit elements as viewed from the foil side of a PC board (refer to Figure 7–2).

   a.

   b.

   c.

   d.

   e.

5. Identify the following components as to their category by placing IMT for "insertion mount technology" or SMT for "surface mount technology" in the blank in front of each.

   _____ **a.** axial-lead resistor   _____ **b.** SOT-23 package

   _____ **c.** SOIC package   _____ **d.** TO-5 package

   _____ **e.** Flat pack   _____ **f.** chip resistor

   _____ **g.** cylindrical   _____ **h.** DIP IC

   _____ **i.** axial-lead diode   _____ **j.** polarized chip

_____ **k.** TO-220 package   _____ **l.** PLCC package

6. Name four types of axial-lead termination methods that can be used on a PC board (refer to Figure 7–12).

   a.

   b.

   c.

   d.

7. List four unacceptable solder joints that you must be able to recognize.

   a.

   b.

   c.

   d.

8. List 10 steps in assembling axial-lead components on a PC board.

   a.

   b.

   c.

   d.

   e.

   f.

   g.

   h.

   i.

   j.

9. List four important points to remember when soldering PC boards.

   a.

   b.

   c.

   d.

10. Place a T for "true" or an F for "false" in front of the following statements for assembling SMT components.

   _____ **a.** A large soldering iron is used on all joints.

   _____ **b.** Pretinning is performed on the component leads and the PC board pads.

   _____ **c.** Flat packs require lead bending before soldering.

   _____ **d.** The reflow technique of soldering is used for SMT components.

_____ e. SMT devices can be held in place with tape while soldering is being done.

_____ f. When soldering a SMT device, until the solder solidifies, the device can be held in place with a pair of tweezers or a small tool with a slight downward pressure.

_____ g. Heat should be removed from a SMT device when you see the solder flow from the pad up onto the lead of the device.

_____ h. Conductive adhesives cannot be used on SMCs.

_____ i. SMT solder joints are acceptable even if the leads of the package are not centered and overhang the edge of the pad.

_____ j. The advantage of using a special heating tool that contacts all leads of a SMT device simultaneously is an even flow of solder, which results in an evenly mounted package.

_____ k. Reflow solder flux is the same as regular solder flux.

_____ l. In some cases, heat can be applied to an SMT device up to 10 seconds without damaging it.

11. Fill-in the blanks in the following statements, making them correct, by using the following words: _circular, same, vacuum, conformal, leads, back-and-forth, soldering iron._

   a. Simple component replacement involves attaching the leads of the new component to the existing _____ of the old component, which are still mounted to the PC board.

   b. Many PC boards have a _____ coating that must be removed before the components can be desoldered.

   c. A desoldering tool having a _____ is used to remove solder from joints.

   d. When replacing new components on a PC board, you would use the _____ techniques required in the original operation.

   e. When desoldering a round lead, you move the desoldering tool in a _____ motion.

   g. When desoldering a wire from a terminal, you first remove the solder with a desoldering tool and then use a regular _____ _____ on the join as you gently shake the wire.

≡≡≡ SECTION 7–4
EXPERIMENTS

EXPERIMENT 1. Soldering Axial-Lead Components onto PC Boards

Objective:

To develop the skills required for good-quality soldering and highly reliable solder joints of components mounted on PC boards.

Introduction:

In this experiment you will practice the skills required to mount various components correctly onto PC boards. Mount and solder each component in the order given: resistors, capacitors, and transistors.

Materials Needed:

1 Pair of diagonal cutters
1 Pair of long nose pliers
1 Short bristle brush
1 Small container of alcohol or other suitable solvent
1 50-watt (maximum) soldering iron
1 Soldering holder
1 Moist sponge
1 Container of flux
1 Small PC board soldering vice
1 Solder-free single- or double-sided PC board
1 Suitable lamp that can be adjusted to shine on the work area
  Some 60/40 rosin-core solder (No. 22)
  Assorted resistors from $\frac{1}{4}$ to 1 W
  Assorted capacitors
  Assorted transistors

Procedure:

1. Examine the PC board to determine where the components will be placed. (Your instructor may have a pre-designated sheet to follow.)
2. Clean the component leads.
3. Clean the terminal pads on the PC board.
4. Measure the distance between the holes on the board where the component is to be placed.
5. Form the leads on the components (see Section 7–1b.1).
6. Insert the component leads into the board holes from the component side.
7. Holding the component in place, turn the board over and trim the leads to the proper length.
8. Clinch the leads in the proper direction.
9. Place the PC board into the vise and solder the leads to the respective pads.
10. Clean the solder joints with solvent.
11. Inspect the solder joints.
12. Compare the solder joints with Figure 7–16.
13. You may also want to mount some resistors or capacitors in a vertical position, as shown in Figure 7–18 (refer to Section 7–1b.5 for more information).
14. After you have completed the mounting and soldering experiment, unplug the soldering iron so that it will cool down while you are cleaning up your work area. Remember to keep the work area as neat and organized as possible.

Fill-in Questions:

1. Before soldering, component leads should be cleaned with a _____ abrasive stick.

2. The time the soldering iron tip is left on the component lead and pad should be no longer than _____ seconds.

3. Unacceptable solder joints would include _____ , _____ , _____ , and _____ .

EXPERIMENT 2. Soldering ICs onto PC Boards

Objective:

To develop the skills required to mount and solder ICs properly onto PC boards.

Introduction:

In this experiment you will solder several ICs onto a PC board. The techniques are similar to mounting axial lead components as far as the soldering process is concerned (refer to Section 7–1c.2 for more specific information).

Materials Needed:

The same tools and PC board as in Experiment 1
Assorted DIP ICs and ICs mounted in TO-5 packages

Procedure:

1. Clean the pads on the PC board where the IC will be mounted with a rubber abrasive stick.
2. If the IC is mounted in a TO-5 package, form the leads so that they will fit into the holes in the PC board (see Section 7–1c.1 and Figure 7–20).
3. If the IC is a DIP, form the pins along each side so that the IC can fit easily into the holes on the board (see Section 7–1c.2 and Figure 7–22).
4. After the DIP IC is inserted into the board, hold the IC in the board as you turn it over and clinch one lead on each side to secure it to the board.
5. Solder the DIP IC to the board.
6. Clean the solder joints with a solvent.
7. Inspect your solder joints.
8. Compare your solder joints with the ones shown in Figure 7–21 or 7–23.

Fill-in Questions:

1. To have an acceptable solder joint on a DIP IC, the solder must be allowed to

flow _____ the hole.

2. After the flow-through, a little more

_____ is added to the joint to mount it up again.

3. The maximum dwell time for soldering

a DIP IC joint is _____ seconds.

EXPERIMENT 3. Soldering Surface Mount Components

Objective:

To develop skills in soldering SMC and in understanding the solder reflow process.

Introduction:

The components used in SMT are very small and difficult to handle. Extreme care must be used at all times.

Materials Needed:

The same soldering tools used in Experiment 1, except that the soldering iron tip must be a very small needle type
Assorted SMT devices
1 Magnifying glass

Procedure:

1. Clean the conductive surfaces of the board with a rubber abrasive stick.
2. Very carefully clean the component leads with alcohol or a similar solvent.
3. Solder the SMT components to the board using the procedures described in Section 7–d. You may use conductive adhesives or the pretin and reflow solder method.
4. Clean the solder joints.
5. Using a magnifying glass, inspect the solder joints for quality.

Fill-in Questions:

1. The reflow method of soldering is ac-

complished by first _____ the component leads and the PC board

_____ . The components are placed on the board and held in place

while _____ is applied, which melts the pretinned solder. The heating device is removed, and when the solder

_____ and solidifies, the component is secure.

2. The SMT components are so small that they may actually stand on end during the soldering process, which is referred

to as the _____ and _____

_____ effect.

- A *PC board* consists of an electrically insulated *substrate,* also called a *laminate,* with a copper foil called the *cladding.*
- The *circuit* is produced on the copper side of the board by using a resist material, photosensitizing, and etching in an acid-type solution.
- The large areas of remaining copper foil are for heat sinks, with the other copper lines being ground and voltage planes, terminals or pads, conductors or runs, and edge connectors.
- An *unsupported hole* in a PC board has no lining and is simply drilled through the substrate.
- A *supported hole* has a metal lining running through the board and may be called a *reinforced hole, through-hole,* or *through-connection.*
- *Eyelets* or *funnelets* may also be used to reinforce a hole.
- A *single-sided board* has the copper circuit on one side only.
- A *double-sided board* has the copper circuit on both sides.
- A *multilayer board* has several conductive layers through it as well as on each side.
- A *flexible board* is a fine-rolled copper laminate circuit bonded to a flexible base of plastic material.
- *Clinched* means that a component lead is bent to one side after it passes through a hole in the PC board, to hold it in place.
- *Swaged* or *spaded* means that the component lead is flattened or bugled after it passes through the hole, to hold it in place.
- *Axial-lead components* can be mounted in an upright position on PC boards.
- When *soldering,* clean component leads, terminals, and/or PC board pads; use solder flux; keep the dwell time short; clean the solder joint with a solvent; and inspect the joint.
- Component leads and pads on PC boards can be cleaned with a *rubber abrasive stick.*
- A *rosin joint* indicates that too little heat was applied to the joint and there is still a quantity of solidified flux between the wire and terminal.
- A *cold solder joint* appears not to cover smoothly and is a result of withdrawing the iron too soon.
- A *disturbed joint* will appear frosty and granulated, indicating that it was moved during solder solidification.
- An *overheated joint* will appear chalky, dull, or crystalline, indicating that too much heat was used.
- *Inspect* all solder joints by rotating them under an overhead light.
- *Multilead components* must be place in the board very carefully so that each lead is in the proper place.
- *Insertion mount technology* (IMT) requires placing component leads through holes in the board to mount them.
- *Surface mount technology* (SMT) involves components that are mounted on one side of the board without going through holes.
- *SMD* stands for "surface mountable device."
- *SMC* stands for "surface mountable component."
- The *reflow solder technique* involves pretinning the pads of the PC

board, applying flux, placing the component on the pads, and heating until the solder wetts the joint.

- A *flat pack* is a surface-mountable IC.
- *Planar mounted* means that all the leads of a component connect on the same side or plane.
- *SOT* means "small outline transistor."
- *SOIC* means "small outline integrated circuit."
- *PLCC* means "plastic-leaded chip carrier."
- *LCCC* means "leadless ceramic chip carrier."
- *Chip components* are SMCs consisting of resistors, capacitors, and inductors.
- A *gull-wing lead* comes out of the package bent downward about 30° and then further on is bent outward 60° to form a gull-wing-looking arrangement.
- The *J-lead* is a lead that is rolled up under the package in the form of a J.
- A *lap flow tool,* a heating device controlled by a foot switch, is excellent for the reflow soldering technique.
- *Reflow solder flux* has microscopic particles of solder suspended in the flux.
- *SMDs must be held in place* with a small tool while being soldered. Masking tape can help to hold them in place.
- *Reflow soldering techniques* are used with SMDs.
- *Electrically conductive adhesives* have a conducting powder of gold, silver, copper, nickel, carbon, or graphite suspended in a bonding material. The adhesive bonds circuits together physically and electrically.
- *Thermosetting adhesives* bond a joint together with the application of a catalyst and cannot be reworked unless the bond is shattered or removed with a solvent.
- *Thermoplastic adhesives* bond a joint together with the application of heat, but can be removed and reworked more easily with the reapplication of heat.
- *Conductive inks* are used to connect prototype circuits and by hobbyists.
- *Sockets* are sometimes mounted on PC boards and are used to hold various transistors and IC packages. Sockets are not as critical to molten solder as the actual components.
- *Simple component replacement* on a PC board may be done by cutting the damaged component from the leads, leaving them in the board. The new component is attached and soldered to the old leads in the board.
- The *conformal covering* may have to be removed from some PC boards before component replacement can be performed.
- A *desoldering tool* with a vacuum is excellent for removing solder from components on a PC board. The tool is placed on the joint and when the solder becomes molten it is drawn into the tool by the vacuum. A circular motion is used with round leads and a back-and-forth motion is used with flat leads.
- Components are *replaced* on a PC board with the same techniques as those used for the original part.

Circle the most correct answer for each question.

1. The base material on which a printed circuit is constructed is called the:

   a. cladding   b. substrate

   c. etchant   d. none of the above

2. The process of producing a PC board where the copper is removed is called:

   a. cladding   b. swaging

   c. etching   d. none of the above

3. The large portions of foil remaining on a PC board are usually:

   a. heat sinks   b. ground or voltage planes

   c. terminals (pads)   d. conductors (runs)

4. A supported hole in a PC board:

   a. is always used with a double-sided board

   b. has a thin metal lining

   c. always has an eyelet

   d. all of the above

5. PC boards can be single-sided, double-sided, or multilayer types.

   a. True   b. False

6. A component lead placed through a hole in the PC board and then bent over onto the foil is said to be:

   a. swaged   b. spaded

   c. clinched   d. all of the above

7. A solder joint that appears frosty and granulated is called a:

   a. rosin joint   b. cold solder joint

   c. overheated joint   d. disturbed joint

8. An acceptable solder joint for a clinched axial-lead component appears with:

   a. just enough solder to fill the hole and some of the lead

   b. solder over the entire clinched lead but with the outline of the lead still visible

   c. enough solder so that the lead is not visible

   d. all of the above

9. A solder joint where the solder appears not to cover the joint smoothly is referred to as a:

   a. rosin joint   b. cold solder joint

   c. overheated joint   d. disturbed joint

10. Axial-lead components can be mounted on a PC board in an upright position similar to radial-lead components.

    a. True   b. False

11. Active components such as diodes, transistors, and ICs are sensitive to heat, and the soldering iron should not be left on the joint longer than:

    a. 1 second   b. 2 seconds

    c. 3 seconds   d. 5 seconds

12. Sometimes a device is placed on the leads of active components to reduce the effect of heat traveling up the lead and is referred to as a:

    a. solder wick   b. crimping tool

    c. heat sink   d. diagonal cutter

13. When soldering diodes and transistors into the holes of a PC board, you must be sure to place the correct lead into the proper hole.

    a. True   b. False

14. A small circle of holes on a PC board is usually for mounting a:

    a. diode   b. DIP IC package

    c. capacitor   d. TO-5 package

15. When soldering a DIP IC, enough solder must be applied so that the solder will flow through the hole and then mount up a little on the soldering side of the PC board.

    a. True   **b.** False

16. When soldering ICs it is important to solder each lead in consecutive order (one after another) so that the general area is kept warm from one soldering joint to the next.

    a. True   **b.** False

17. Small components whose leads do not fit into holes, but are only held in place by the solder are referred to as:

    a. edge connectors   **b.** insertion mount technology

    c. printed circuitry   **d.** surface mount technology

18. One difference between IMT and SMT is:

    a. SMT devices cannot be soldered with a soldering iron.

    b. SMT devices require more room on the PC board for comparable circuits than do IMT devices.

    c. SMT devices require more heat sink area.

    d. SMT devices use a special flux that contains powered solder.

19. Using the solder reflow method:

    a. the iron is placed on the joint and solder is placed on the iron. The solder liquefies and runs down to the joint.

    b. the component leads and PC board pads are pretinned, the component is put in place on the board, and heat is applied.

    c. both of the above are acceptable.

    d. neither a nor b is acceptable.

20. One of the most important points to remember when soldering PC boards is to:

    a. place flux on all component leads before soldering to ensure a good-quality solder joint

    b. apply plenty of pressure with the soldering iron on the PC board so that there is good heat transfer, making the process of soldering easier

    c. gently rest the soldering iron on the PC board and keep the dwell time as short as possible, to form a good-quality solder joint

    d. all of the above are acceptable.

ANSWERS TO FILL-IN QUESTIONS AND SELF-CHECKING QUIZ

Experiment 1: (1) rubber   (2) two   (3) rosin joint, cold solder joint, disturbed joint, overheated joint

Experiment 2: (1) through   (2) solder   (3) three

Experiment 3: (1) pretinning, pads, heat, cools   (2) drawbridge, tombstome

Self-Checking Quiz: (1) b   (2) c   (3) a   (4) b   (5) a   (6) c   (7) d   (8) b   (9) b   (10) a   (11) b   (12) c   (13) a   (14) d   (15) a   (16) b   (17) d   (18) d   (19) b   (20) c

Unit **8**

WIRING PROCEDURES

INTRODUCTION   In the past, copper wires connected all the components in an electronic system. The PC board has replaced many wires in the system with copper circuit patterns. However, PC boards are usually connected together by wire cables or harnesses. There are several techniques used to wire PC boards, front-panel switches, controls, and other devices together into a system.

UNIT OBJECTIVES   Upon completion of this unit, you will be able to:

1. Define *cable, harness, lacing, wire wrapping techniques, flat-ribbon cable,* and other terms associated with wiring procedures.
2. Make spot ties on a cable.
3. Tie wires together into a harness with lacing cord.
4. Describe the use of a harness board.
5. Install nylon self-locking straps on bundles of wire.
6. Assemble connectors and sockets to wires.
7. Demonstrate how to prepare a shielded cable for assembly.
8. Explain the proper use of wire wrapping tools and procedures.
9. Strip flat-wire cable.
10. Assemble flat-wire cable connectors.

≡≡≡ SECTION 8-1
FUNDAMENTAL CONCEPTS

8-1a WIRE CABLES

A *wire cable* constitutes several wires of the same length contained in a single electronically insulated housing or case. The inner wires are also insulated and may go to a connector, switch, or other device. Many types of cables with wires of various sizes in varying numbers can be purchased. Figure 1-5g, h, and i in Unit 1 shows some simple cables. Often, cables with a specific number of wires and sizes cannot be bought and the manufacturer must make the cables for the equipment.

8-1a.1 Assembling Wire Cables

Figure 8-1 shows various ways in which cables can be assembled. A group of wires may simply be placed into an insulated case (Figure 8-1a). The case may be tied at each end to keep it from sliding back and forth (Figure 8-1b). There are cases that have a zipper arrangement where the wires are placed inside and the zipper closed (Figure 8-1c). Large heat-shrinkable tubing can also be used to make cables. The wires are placed inside the tube and then a large heating gun is used to shrink the tubing about the wires (Figure 8-1d). Nylon (plastic) ties are used to form cables (Figure 8-1e). An older method of making cables involves tying the wires with linen cord (Figure 8-1f).

The basic tools used for making cable ties are shown in Figure 8-2. The *linen cord* used for tying cables is usually impregnated with wax to keep it from slipping on the wires. The nylon tie is a long and slender strap with one end pointed, and the other end has a collar with a slit. One side of the tie is serrated. The nylon tie tool is used to pull the tie tightly on the wire.

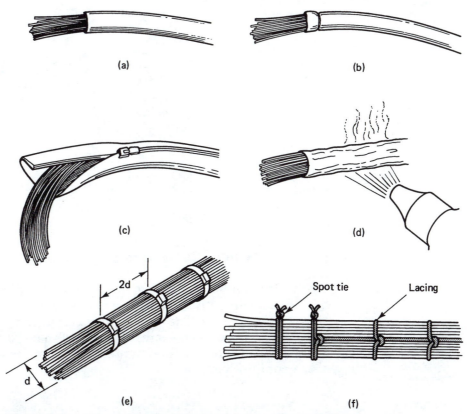

Figure 8–1 Types of cables (a) loose; (b) end-tied; (c) zipper; (d) shrinkable tubing; (e) nylon cable tie; (f) linen tie and lacing.

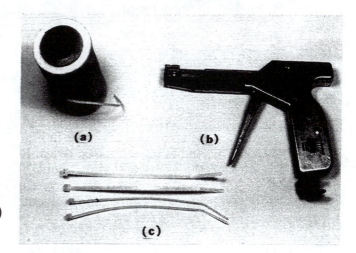

Figure 8–2 Cable-tying devices: (a) linen cord; (b) nylon tie tool; (c) nylon ties.

8–1a.2 Using the Nylon Self-Locking Tie

The tie is placed through the slit, forming a loop around the bundle of wires. The serrated side must be facing the wires. The strap is slid into a groove on the tool. When the handle is squeezed on the tool, the strap is tightened around the bundle of wires. The strap is pulled through the slit and locks itself on the serrations. The strap is then cut with the tool, or it is removed and the remaining strap is cut off with a pair of diagonal cutters. Care must be taken not to leave a tie too loose or so tight that the wires are overly cramped and could become damaged. Figure 8–3 shows a nylon self-locking tie being tightened with a tool. Some self-locking ties have a bracket for securing the cable to the chassis or panel.

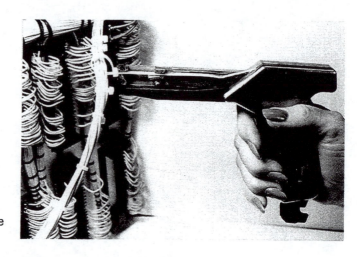

Figure 8–3 Nylon spot tie using tool.

8–1a.3 Tying a Cable with Linen Cord

Linen cord can be used to produce *spot ties* (individualized ties) along a bundle of wires, thus creating a cable. Figure 8–4 shows one method of making a spot tie. The cord is placed around the wires, crossed over, and placed around them again. The loops of the cord are then slid together before the ends are pulled to tighten the tie on the wires. The loose ends of the cord are then cut.

Figure 8–4 Making a spot tie: (a) over and under; (b) around and back under; (c) loops are apart; (d) slide loops closer together and pull cord in opposite directions.

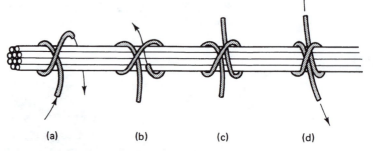

(a)   (b)   (c)   (d)

Another method of making a spot tie is a simple loop and then a square knot, as shown in Figure 8–5. The cord is placed around the wires and is then knotted twice. Remember not to leave the cord too loose or the wires will slide about. Also, if the tie is too tight, the wires could be strained and cause trouble. Make the ties fairly snug so that the wires beneath cannot move about very easily.

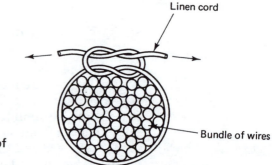

Linen cord

Bundle of wires

Figure 8–5 Endview of tying a square knot.

Cable lacing is the technique of producing many ties on a bundle of wires with a single length of linen cord. Figure 8–6 shows the technique for lacing a cable. A spot tie is first placed on the bundle of wires, leaving one end of the cord very long. This long end of the cord is then

Figure 8–6 Lacing a cable.

used to make continuous ties along the bundle of wires. A small loop is made in the cord as it is placed around the wires. The long end is then pulled through the loop to tighten the cord on the wires. Each tie should be snug and self-locking or the entire lacing may slip and be of no use. All ties should be along a straight line to form an efficient cable. When the end of the lacing is reached, the cord is wound around the wires and a knot is formed to seal the lace.

8–1b THE WIRE HARNESS

Unlike a wire cable, where all the wires are the same length, the *wire harness* has wires of varying lengths that connect to various components in a system.

8–1b.1 Typical Harness

A typical wire harness is shown in Figure 8–7. This harness connects together various PC boards, a power supply (lower left), and cable connectors (lower right). The wires begin at one place or termination and then go into a common bundle. The wires "break out" at various places from the common bundle to connect to the other termination points. If the wiring is for a single unit, each wire is routed from one terminal point to the other point. After all wires are routed and soldered in place, they are tied into a harness. Most often, though, many units requiring the same harness will be constructed. In this case, the *prototype* (the first one) is not permanently soldered in place, but is completed and the wires are laced into a harness. The connections are then unsoldered and the harness is removed. This prototype harness is then used as a *template* (model for making other harnesses).

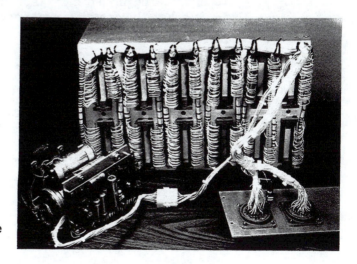

Figure 8–7 Typical cable and harness application.

The harness is laid on a large piece of drafting paper and an outline of it is drawn. The outline is then attached to a wooden board, referred to as a *wire-harness jig board* or simply a *harness board*. Pegs and/or springs are mounted on the board where the harness will lie and where breakouts occur. Some components, such as switches, cable connectors, and terminal strips, may be attached to the harness board as shown in

Figure 8–8. The harness should be neat and orderly, with all wires running parallel to each other. Wires should not weave in and out or cross over each other unnecessarily, and ties should be spaced evenly.

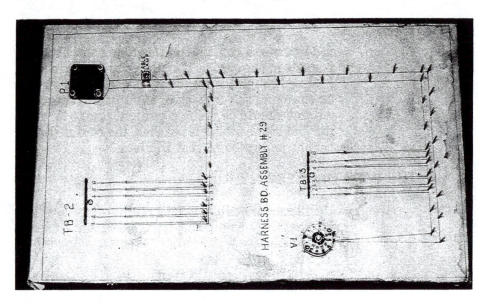

Figure 8–8 Harness wiring board.

8–1c CABLE CONNECTORS

Numerous cable connectors are used in the electronics field. Figure 1–8 introduced you to coaxial, pin-contact, and rack-and-panel connectors. Connectors may use round pins, square pins, flat-blade type, or other configurations to make contact.

Cable connectors require a matching set. Each connector has a mate. The connector with pins is referred to as the *male,* and the connector with holes or jacks is called the *female.* To ensure proper alignment and connection with each other, the connectors will have a *key* in the form of an offset pin, slot, or other distinguishing feature. Some connectors attach to the chassis permanently and can be identified by a mounting flange with holes.

8–1c.1 Solder Connector

Some *solder connectors* have the wires soldered to lug-type terminals. Figure 8–9 shows a solder lug called a D-sub connector. This connector may have gold-plated contacts that connect to its mating unit. Refer to Section 6–1b.4 and Figure 6–9 for the soldering procedures for this type of terminal. The terminal pins are identified by letters and/or numbers on the connector. In some cases, heat-shrinkable tubing is placed on the

Figure 8–9 Solder-type edge connector.

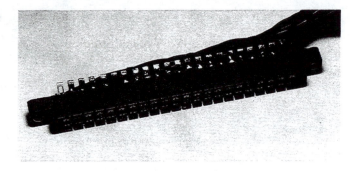

wires before soldering, and then after the joint has solidified the tubing is pushed over the connection and heated to form an insulated terminal.

8–1c.2 Snap-in Pin Connector

The *snap-in pin connector* has its terminals crimped on the wires first, and then these terminals are snapped into the connector housing. Figure 8–10 shows a snap-in pin connector. Notice the terminals crimped to the wires not yet placed into the housing. A special tool is required to remove the terminals once they are in the housing.

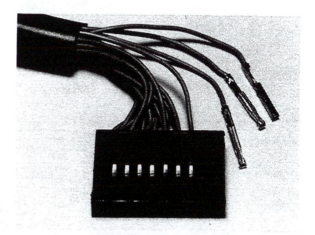

Figure 8–10 Snap-in pin connector.

8–1c.3 AMP Cable Connectors with Multiple Parts

Some connectors consist of several parts that must be assembled after the wires are soldered in place. The *AMP cable connector* shown, in Figure 8–11, is designed for commercial applications such as automotive, aircraft, instrumentation, computer, and peripheral equipment. This type of connector offers the features of a cable clamp, which prevents excessive strain on the cable or wires from reaching the contacts inside the connector, and also provides a permanent connection to its mating unit by having a threaded collar.

AMP connectors use cup terminals for securing the wires. Refer to Section 6–1c.3 and Figure 6–24 for the soldering procedures of cup terminals. The terminal pins are identified by letters and/or numbers on the connector.

Figure 8–11 Amphenol connector: (a) unassembled; (b) assembled.

(a)

(b)

8–1c.4   Flat (Ribbon) Cable Connector

Flat wire or *ribbon cable* is composed of insulated wires bound together side by side to form a flat conductor. Ribbon cable is inexpensive to manufacture, lightweight, flexible, and relatively easy to use. It can be used in equipment of varying size and has found widespread application in the computer industry.

Most ribbon cable is attached to connectors by having their wires pierced by pointed contacts, as shown in Figure 8–12. The cable is cut

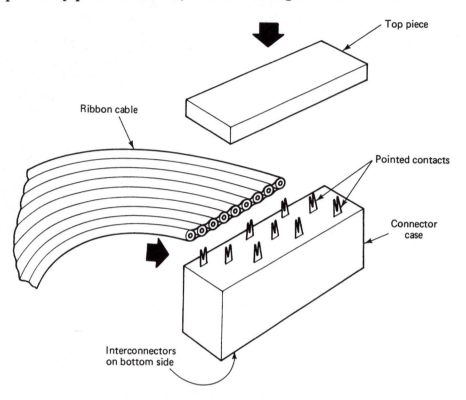

Figure 8–12   Ribbon cable connector assembly.

evenly along the end to be mounted. The cable is placed over the pointed contacts and then the top piece of the connector is pushed into place, which causes the wires to be pierced by the contacts. The top has groves or tabs that lock it in place. Other attachments to the connector can provide strain relief to the ribbon cable. Figure 8–13 shows a completed ribbon cable connector.

In cases involving the soldering of ribbon cable, the wires must be stripped. One method is with a special wide blade that is used in a hand-plane manner to remove the insulation. Another method is dual buffing wheels, which contact both sides of the cable and are used to heat the insulation by friction. The insulation is then pulled from the wires. The insulation can also be cut from ribbon cable using a special heated knife.

Figure 8–13   Flat ''ribbon'' cable connector.

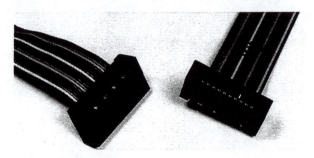

The heated knife is pushed down on the cable and melts a groove in the insulation; then the cable is pulled from beneath the knife, leaving the exposed wire.

8–1c.5 Shielded-Cable Connections

Shielded cable is used for many wire connections, such as phono/phone plugs and jacks. *Shielded-cable metallic braid* serves as a conductor in two-wire cable systems. The plugs in phono/phone cable are manufactured on the cable by crimping. When shielded cable has to be soldered to plugs or jacks, special preparation is required, which must be carried out as shown in Figure 8–14.

In cases where the metal braid is the outer layer of the cable, the braid must be separated with a pointed tool (Figure 8–14a). You must be very careful not to break any strands of the braid. These loose or open strands are then twisted into a wire that is easier to solder (Figure 8–14b).

If the metal braid is between the inner and outer insulation of a shielded cable, the *poke-through method* is used. The outer insulation is removed and the braid is pushed backward, exposing the inner insulation and conductor (Figure 8–14c). The pointed tool is used carefully to

Figure 8–14 Shielded-cable preparation: (a) open braid with pointed tool; (b) twist braid to form ground connection; (c) push braid back on insulation and open hole with pointed tool; (d) pull insulation through hole; (e) trim insulation from inner conductor.

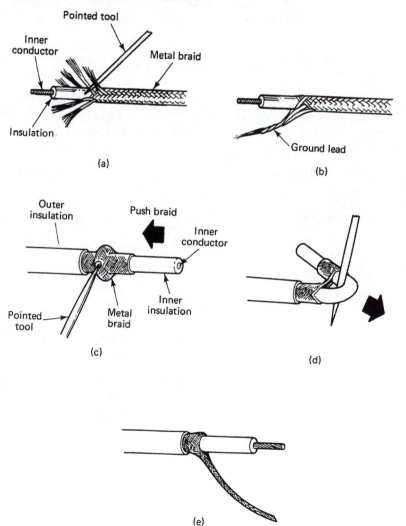

open a hole in the braid. Once sufficient braid has been separated from the cable, the inner conductor is pulled through the hole (Figure 8–14d). The loose braid is then twisted to form the other conductor. The inner insulation is stripped away, exposing the inner conductor (Figure 8–14e), and the cable is ready for soldering.

8–1d   WIRE WRAPPING TECHNIQUES

Wire wrapping is the technique of connecting wires to terminals without soldering. A connection is made by tightly wrapping wire around a sturdy metal terminal, sometimes referred to as a *stud*, post, or pin, which is mechanically and electrically acceptable.

8–1d.1   Wire Wrapping Materials and Tools

The stud used to wrap the wire around is usually made of a copper alloy and is square or rectangular, as shown in Figure 8–15. The stud may be 0.25 to 1 in. or more in length with side dimensions of 0.025 to 0.45 in. The wire used for wire wrapping is single stranded and ranges in size from No. 20 to No. 30.

A special *wire wrapping tool* is needed to wrap the wire around the stud. The nozzle of the tool may be operated manually or by automatic tools, including a pneumatic (air pressure) motor or electric motor. Automated machines have been used for many years that measure the wire, strip it, and place it on predetermined studs under the control of a computer program. A *wire de-wrapping tool* is used to remove wires for repairing.

8–1d.2   Wire Wrapping Procedures

Wire wrapping must be done carefully to achieve good results. Practice will improve your ability to produce good wire wraps. The following procedure should aid you in wire wrapping.

Figure 8–15   Wire wrap terminals.

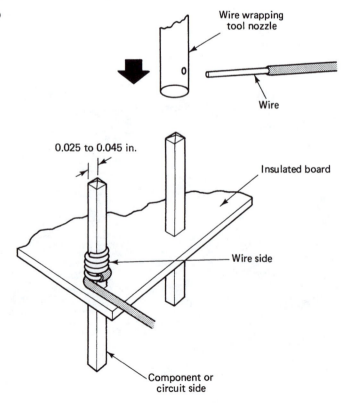

1. Remove sufficient insulation from the wire so that at least five turns of wire can be made on the stud. Remember not to nick the wire, or the strength of the connection may be weakened.

2. Insert the stripped wire into the hole of the tool (with automatic tools the wire is usually fed to the tool from a spool). Place the tool over the stud.

3. Wrap the wire on the stud according to the type of tool you are using. The wire should be wrapped tightly and neatly around the terminal as shown in Figure 8–16. There are two types of acceptable wire wraps: class 1, where no insulation touches the stud (Figure 8–16a), and class 2, where one turn of insulation is made on the stud (Figure 8–16b).

Figure 8–16 Acceptable wire wraps: (a) class 1 with no insulation on terminal; (b) class 2 with insulation on terminal.

(a)

(b)

4. After the wrap is finished, inspect it for acceptance. Figure 8–17 shows three unacceptable wire wraps.The wrap should not have too much space between the turns of wire on the stud (Figure 8–17a). The wrap shuld be symmetrical or even on the stud and not have an open (Figure 8–17b). Wires should not cross over each other on the stud (Figure 8–17c). Very often, DIP IC sockets manufactured with wire wrapping posts are used in complex PC boards where there are a large number of ICs.

Figure 8–17 Unacceptable wire wraps: (a) spiral wrap—too much space; (b) open wrap—not symmetrical; (c) over wrap—wire crosses over.

(a)

(b)

(c)

≡ SECTION 8–2
DEFINITION EXERCISE

Write a brief description of each of the following terms.

1. Wire cable _____

_____

2. Loose tubing _____

_____

3. Shrinkable tubing _____

_____

4. Zipper tubing _____

_____

5. Nylon cable ties _____

_____

6. Linen cord _____

_____

7. Spot ties _____

_____

8. Cable lacing _____

_____

9. Wire harness _____

_____

10. Breakout _____

_____

11. Prototype _____

_____

12. Template _____

_____

13. Wire-harness jig board _____

_____

14. Harness board _____

_____

15. Male connector _____

_____

16. Female connector _____

_____

17. Key _____

_____

18. Solder lug connector _____

_____

19. Snap-in pin connector _____

_____

20. AMP cable connector _____

_____

21. Flat wire or ribbon cable _____

_____

22. Shielded-cable metallic braid _____

_____

23. Wire wrapping _____

_____

24. Wire wrapping tool _____

_____

25. Wire de-wrapping tool _____

_____

≡ SECTION 8–3
EXERCISES AND PROBLEMS

Complete this section before beginning the next section.

1. List six types of cables that might be assembled by an electronics firm.

   a.   **b.**

   c.   **d.**

   e.   **f.**

2. Match the descriptions in column A with the correct device in column B.

Column A	*Column B*
_____ **a.** Pins or plugs of a connector	**1.** cable
	2. harness
_____ **b.** A linen cord tie at one place	**3.** Spot tie
	4. lacing
_____ **c.** A group of combined wires going from one location to another location	**5.** harness board
	6. key

_____ **d.** A jig board upon which wires are assembled

_____ **e.** The part of a connector that ensures proper alignment

_____ **f.** Holes or jacks of a connector

_____ **g.** A wire connection without solder, crimping, or screws

_____ **h.** A wire with metallic braid

_____ **i.** Continuous linen cord ties

_____ **j.** A group of wires with many breakouts to various terminals

7. male connector
8. female connector
9. wire wrapping
10. shielded cable

3. List four types of cable connectors.

   a.   **b.**

   c.   **d.**

4. Briefly describe how to prepare shielded cable for soldering.

5. Describe how to assemble the most common type of connector used with ribbon cable.

6. Give a brief description of how to perform wire wrapping.

≡≡≡**SECTION 8-4**
EXPERIMENTS

EXPERIMENT 1. Making a Cable with Sleeving

Objective:

To develop skills in making cables.

Introduction:

A cable is used to combine the wires of a circuit or system into a neat identifiable form. A cable keeps wires from interfering with other components of the circuit, from hooking or catching on things, and provides a measure of safety.

Materials Needed:

1 Piece of sleeving 2 to 3 ft long
1 A bundle of wires long enough to extend beyond the sleeving 2 to 3 in. at each end
1 Roll of linen lacing cord
1 Pair of diagonal cutters
1 Container of talcum powder

Procedure:

1. Make sure that there are enough wires to fill the center of the sleeving but not to become too tight.
2. Tie the end of the bundle with the lacing cord, leaving enough on one end of the cord to pass all the way through the sleeving.
3. Lightly sprinkle some talcum powder on the bundle of wires and spread it with your hands so that it forms a light film on the wires. This will make the bundle of wires slide more easily through the sleeving.
4. Feed the lacing cord tied to the bundle through the sleeve.
5. Pull the bundle of wires through the sleeve until about 3 in. shows at the end.
6. Make a spot tie with the lacing cord about 2 in. from each end of the sleeving.
7. Trim excess cord ends with the diagonal cutters and remove the cord from the bundle of wires.

Fill-in Questions

1. A cable is used to keep a group of wires

   _____ and orderly.

2. Talcum powder is used on the wires so

   they will _____ more easily
   through the sleeving.

EXPERIMENT 2. Lacing with Linen Cord

Objective:

To develop the skills required for lacing a cable or harnass.

Introduction:

This experiment can be a simple cable, or you can make a wire harness if a harness board is available.

Materials Needed:

1 Bundle of wires
1 Roll of linen lacing cord
1 Pair of diagonal cutters

Procedure:

1. Assemble the wires into a bundle or place them in correct order on the harness board.
2. Make a spot tie at one end of the bundle of wires. Leave one end of the cord about two to three times longer than the length of the wires.
3. Begin the lacing technique as described in Section 8–1a.3 and shown in Figure 8–6. Remember to tighten each tie and leave the proper space between the ties. Continue this procedure until you reach about 1 in. from the other end of the bundle of wires.
4. Make a sealing knot with the cord at the end of the bundle of wires and cut the remaining cord with the diagonal cutters.
5. Inspect your lace and make sure that it is straight and all ties are tight.

Fill-in Questions:

1. To begin lacing a bundle of wires, a

   _____ _____ is first performed.

2. The ties in a lace should be _____

   _____ and _____ .

3. A lace is finished with a _____ knot at the end of the bundle of wires.

EXPERIMENT 3. Assembling Cable Connectors

Objective:

To become familiar with various cable connectors.

Introduction:

There are several types of cable connectors

which are soldered to wires, although the soldering techniques used are the same as with the individual terminals studied previously.

Materials Needed:

1 Wire stripper
1 Pair of diagonal cutters
1 Pair of long nose pliers
1 50-watt (maximum) soldering iron
1 Soldering iron holder
1 Moist sponge
1 Container of flux
1 Container of solvent or alcohol
1 Small stiff bristle brush
1 Small soldering vise
1 Appropriate wires for soldering connectors or cable
1 Ribbon cable and connector
1 Piece of shielded cable
Various solder-type cable connectors
Some 60/40 rosin-core solder (No. 22)

Procedure:

1. Set up your work area and heat the soldering iron. Use the procedure for tinning a soldering iron described in Section 6–1b.3.
2. Place the cable connector into the vise.
3. Refer to Section 6–1b.8 for tinning of the wires in the cable.
4. Refer to the appropriate section depending on the type of terminals on the connector: lug type, see Section 6–1b.4 and Figure 6–9; cup terminal, see Section 6–1c.3 and Figure 6–24.
5. Be sure to clean and inspect each solder joint.
6. Trim a piece of ribbon cable and assemble the connector as described in Section 8–1c.4 and shown in Figure 8–12.
7. Prepare a shielded cable for soldering as given in Section 8–1c.5 and shown in Figure 8–14.

Fill-in Questions:

1. The same procedures for soldering ___

   _____ terminals usually apply to soldering cable connectors.

2. Some cable connectors require varied

_____ to assemble.

EXPERIMENT 4. Wire Wrapping Techniques

Objective:

To develop skills in the proper method of wire wrapping.

Introduction:

Wire wrapping is a solderless method of connecting wires to terminals that is mechanically and electrically acceptable. The wire wrapped around the stud must be even and neat.

Materials Needed:

1 Mounted wire wrapping terminals to a PC board or IC socket
1 Wire wrapping tool
1 No. 20 to No. 30 single-stranded wire
1 Pair of wire strippers
1 Pair of diagonal cutters
1 De-wrapping tool
1 Small vise

Procedure:

1. Place the terminal part into the vise.
2. Using the methods given in Section 8–1d.1, wire wrap several terminals.
3. Inspect your work and refer to Figures 8–16 and 8–17.
4. Terminals that are not acceptable can be rewired using a de-wrapping tool. Place the tool on the terminal and turn it in the direction opposite to that in which the wire is wound. As the wire comes off the terminal, pull on it gently. De-wrapping a terminal can be done to an acceptable terminal for practice.
5. Re-wrap the terminal. Try placing a second wire wrap on the same terminal, above the original wrap.

Fill-in Questions:

1. Wire wrapping is a _____ method of connecting a wire to a terminal.

2. Wire wrapping is a _____ method than soldering of connecting wires to terminals.

≡ SECTION 8–5 INSTANT REVIEW

- A *wire cable* consists of several wires of the same length contained in a single electrically insulated case.
- The *case* on a cable may be loose tubing, shrinkable tubing, zipper tubing, or nylon cable ties and linen cord.
- A *spot tie* is a single tie on a bundle of wires.
- *Nylon cable ties* are spot ties installed with a special tool.
- *Linen cord* is used to make spot ties.
- *Cable lacing* is a continuous group of cable ties made with linen cord.
- Unlike a cable, a *wire harness* has wires of varying lengths that connect to various components in a system.
- *Breakouts* in the harness allow the various terminals to be connected together.
- A *prototype* is a first run or initial model of a piece of equipment.
- A *template* is a pattern for making wire harnesses.
- A *wire harness* or *jig board* is a board with pegs and/or springs on which wires are assembled to produce a harness.
- A *male cable connector* has pins or plugs.
- A *female cable connector* has holes or jacks.
- A *key* on the cable connector allows proper alignment of the two parts of the connectors.

- Wires *attach* to cable connectors by means of solder lugs, cup terminals, and crimp/snap-in types.
- Usually, *flat* or *ribbon cable attaches to connectors* by means of small pointed pins that pierce the insulation and make contact with the wire.
- Special preparation is required to place connectors on *shielded cable.* The braid must be carefully separated and then twisted together to form a conductor.
- *Wire wrapping* is a solderless method for connecting wires to terminal points. Special *wire wrapping tools* are required to place the wire on the studs. A special *de-wrapping tool* is also used to remove the wire from a stud.

SECTION 8–6
SELF-CHECKING QUIZ

Circle the most correct answer for each question.

1. A group of wires, all of the same length, contained in a sleeve is referred to as a:

   a. cable   **b.** harness

   c. both a and b   **d.** neither a nor b

2. One technique *not* used in making a wire cable is:

   a. lacing   **b.** nylon cable ties

   c. zipper tubing   **d.** wire wrapping

3. A group of wires that is laced together and has breakouts is referred to as a:

   a. cable   **b.** harness

   c. cable lacing   **d.** none of the above

4. The first run or initial piece of equipment is called a:

   a. template   **b.** key

   c. prototype   **d.** none of the above

5. In assembling a snap-in cable connector, first the:

   a. wires must be tinned

   b. contacts on the connector must be tinned

   c. contacts must be crimped on the wires

   d. none of the above

6. Special groups of wires that need to be reproduced many times are manufactured with a:

   a. wire wrapping tool   **b.** harness board

   c. key   **d.** all of the above

7. In preparing a shielded cable, the metallic braid must be cut away from the inner conductor.

   a. True   **b.** False

8. To match and join cable connectors properly, you must locate the:

   a. male side   **b.** female side

   c. template   **d.** key

9. When using a wire wrapping tool, the turns of wire on the stud must be wrapped on top of each other.

   a. True   **b.** False

10. To remove a wire from a wire wrapped stud, you would use:

   a. a de-wrapping tool   **b.** diagonal cutters

   c. long nose pliers   **d.** all of the above

ANSWERS TO FILL-IN QUESTIONS
AND SELF-CHECKING QUIZ

Experiment 1: **(1)** neat **(2)** slide

Experiment 2: **(1)** spot tie **(2)** tight, straight **(3)** sealing

Experiment 3: **(1)** individual **(2)** techniques

Experiment 4: **(1)** solderless **(2)** faster

Self-Checking Quiz: **(1)** a **(2)** d **(3)** b **(4)** c **(5)** c **(6)** b **(7)** b **(8)** d
(9) b **(10** a

Unit 9

Repairing Electronic Devices and PC Boards

INTRODUCTION

State of the art is a term coined to express the latest development stage in manufacturing technology. When someone refers to the state of the art regarding a piece of electronic equipment, it means the newest materials, circuits, devices, and procedures in use, with the connotation that the product is better, more efficient, and easier to use. Changes in the electronic industry occur at a phenomenal rate, with more modular assemblies and increased packing density of components. The repair of electronic equipment can be very difficult, and very often specialized training is required for the assembler or technician involved in repair work. Because of the complexity of the repair situation in today's world, in this unit we cover only the basic concepts and general practices of repair.

UNIT OBJECTIVES

Upon completion of this unit, you will be able to:

1. Define the repair process.
2. Explain the proper use of repair equipment.
3. List safety precautions in repairing electronic components and devices.
4. Explain how to repair obvious problems.
5. Show the procedures for replacing components on a PC board.
6. Explain how to repair a damaged PC board.

9–1a   THE REPAIR PROCESS

There is more to repairing today's electronic devices than simply replacing components. Persons making the repairs must understand thoroughly the types of materials and components with which they are working. They must understand how to use selected procedures and work constantly with the idea of *quality assurance:* that after repair, the device will function as good or better than the original unit.

9–1a.1   Types of Repair

There are four types of repairs or operations involving similar types of procedures and methods, with each requiring the same quality assurance.

1. *Rework:* reestablishing an assembly's functional and physical characteristics without deviating from the original manufacturing specifications. This is usually performed at the factory level before final inspection and shipment of the product.
2. *Repair:* reestablishing the functionality, reliability, and quality characteristics of an assembly that has failed or has been damaged and may require deviations from the original manufacturing specifications. This type of repair may be done at a service center or returned to the factory.
3. *Modification:* updating an assembly to meet a new or modified specification by eliminating, replacing, and/or adding new components. Work such as this is usually done in the field or at a service center. In some cases the unit will have to be returned to the factory for modifications.
4. *Salvage:* removing parts from a circuit board or assembly, without damage, for reuse in other assemblies. Sometimes PC boards are

salvaged for reuse. Salvage operations are usually performed at the factory.

9–1a.2 Elements of Repair

There are five elements of repair that should be attempted to assure reliable end results that meet original factory quality.

1. *The skill of the repair person:* The person performing the repair must be thoroughly trained in the procedures and methods of the specific repair job.
2. *Proper tools and equipment:* With today's sophisticated assemblies and manufacturing techniques, tools for repairing must be equal in quality to those used in the original manufacturing process.
3. *Analysis and definition of the problem:* The repair job must be analyzed and procedures established, with typical considerations being:
   a. What is the extent of damage, or what components have failed?
   b. Is the PC board single-sided, double-sided, or multilayered?
   c. Is the board conformally coated, and with what type of coating?
   d. Of what base material is the board made?
   e. What are the thermal properties of the circuitry?
   f. Is hole support used in the assembly?
   g. What are the mechanical, thermal, and electrostatic properties of the assembly?
   h. What type of component interconnects are used, and do they need repair?
   i. What types of solder joints are used, and what are the thermal mass properties of each?
4. *Defined procedures of work:* These are specific procedures that must be understood and performable under the specific repair conditions, whether in a factory, service center, or under field conditions. Typical procedures may include:
   a. Removal of conformal coatings on PC boards.
   b. Removal and/or repair of damaged or lifted circuit conductors, such as runs, pads, or through-hole plating.
   c. Removal and replacement of defective components.
   d. Repair and replating of edge connectors.
   e. Modification of circuit boards to meet a new or changed engineering specification, or a changed functional requirement.
5. *Quality assurance:* The repair person must assume responsibility for the end result of the repair job. The piece repaired should have the same quality as when it was manufactured. The repair person has to inspect the finished repair job and know that it complies with the specific repair procedures and the detailed standards before giving the final okay.

9–1a.3 Personal Attributes for Successful Repair

The repair person does not necessarily need to understand the components, device, or system's intended use, its detailed functions, or its manufacturer. It is only necessary to understand and deal effectively with the physical and mechanical properties of the circuit board assembly and its components, and to be able to restore them efficiently to their original capability. This person has total control over the reliability and quality of the operational hardware under repair. The equipment's original reliability can either be maintained or degraded in varying degrees. The repair person must have certain personal attributes that will enhance the success of the repair job:

1. *Skills:* The repair person must be a highly skilled craftsperson with good analytical ability. All aspects of the repair process must be known.
2. *Responsibility:* The person must have pride in the work performed and be responsible for its outcome.
3. *Patience:* Restoring PC boards to their original condition requires a great deal of patience, and each step in the process must be done with accuracy. The person must have sufficient self-control to perform each part of the operation to its fullest.
4. *Confidence:* Confidence is formed by performing good-quality work. Similarly, being confident lends itself to being responsible and developing a reputation for turning out good-quality work.

9–1a.4 Proper Repair Equipment

The most important part of any technical/mechanical job is to have the proper tools in good working order. There should also be a suitable stock of supplies, such as abrasives, solvents, cleaning solutions, spare parts, and miscellaneous items used in specialized work. Repairing electronic PC boards and devices requires various tools and may require special equipment. The following list shows the tools that are essential and those that would be desirable.

1. Basic hand tools
   a. Diagonal cutters
   b. Long nose pliers
   c. Wire stripper
   d. Small set of wrenches
   e. Set of flat-blade screwdrivers
   f. Set of Phillips-head screwdrivers
   g. Set of Allen and Bristo wrenches
   h. Set of various-size nut drivers
   i. Small set of files
2. Soldering tools
   a. Soldering iron
   b. Assortment of various-size soldering tips
   c. Soldering iron holder
   d. Soldering aids
   e. Moist sponge and holder
   f. Clip-on heat sinks
   g. Soldering flux
   h. Rosin-core solder 60/40
3. Desoldering tools
   a. Wicking braid
   b. Desoldering bulb
   c. Spring-loaded desoldering tool
   d. Vacuum desoldering tool
   e. Special soldering tips for DIP ICs and SMT devices

4. Miscellaneous supplies
   a. Alcohol or other suitable solvents
   b. Various epoxies, depending on the types of PC boards to be repaired
   c. Cleaning rags
5. Special equipment
   a. Special repair station with soldering and desoldering irons, constant vacuum system, small grinding accessories, special light with magnifier glass, and other special attachments.
   b. Special equipment for the removal and replacement of SMDs. This type of equipment may have a microscope or closed-circuit television used for accurate alignment of components. A vacuum device is used to place the component in the location. Hot air is generally used to solder the SMDs in place, thereby preventing damage to surrounding components and the board.

Figure 9–1 shows an ideal repair station with all the special features needed to perform complex and sophisticated repair on PC boards.

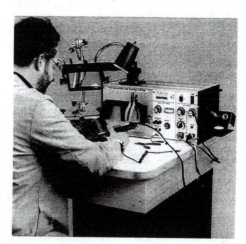

Figure 9–1 Ideal repair station. (Permission to reprint granted by PACE Incorporated, 9893 Brewers Court, Laurel, MD, 20707.)

9–1b REPAIR STATION SAFETY PRECAUTIONS

Safety is the first requirement at any work location. Safety precautions must be taken to protect not only the repair person, but also the equipment being worked on. The following list should be used to establish a safe work location.

1. The workstation should be well organized, with tools and equipment in proper working condition.
2. The work area should be well lighted.
3. The work area should be kept clean and free of debris.
4. Safety glasses or goggles should be worn while work is being performed.
5. Appropriate safety and caution signs should be displayed where there is a potential safety hazard.
6. A prescribed first-aid kit for the type of work being done should be available, including:
   a. Ointments and bandages for cuts and abrasions.
   b. Ointments and bandages for burns.
   c. Eyewash.

Precautions should be taken to protect MOS devices from static electrical charges. Figure 9–2 shows a desirable workstation for MOS devices. The boards have antistatic bags placed over them at all times unless they are being worked on. Short sleeves should be worn by the repair person so that the grounding wrist strap will make a good connection to the skin.

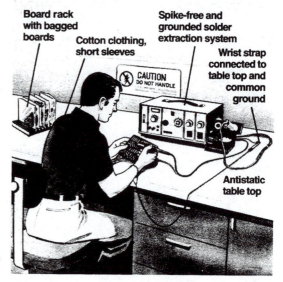

Figure 9–2 Protection against static electrical charges. (Permission to reprint granted by PACE Incorporated, 9893 Brewers Court, Laurel, MD, 20707.)

An antistatic tabletop is recommended. Voltage spike–free and properly grounded work equipment is also required. Electrical overload problems that cause voltage spikes are shown in Figure 9–3.

In a typical work situation (Figure 9–3a), the soldering or desoldering iron is connected to a power supply which, in turn, is connected to the main source of electrical current. Switches, electrical contacts, and motors operating in other parts of the room and building can cause voltage spikes in the line voltage. If a voltage spike occurs in the line voltage, it can pass right through the power supply and soldering tool to the workpiece. If the apparatus causing the voltage spikes cannot be relocated, a line filter (Figure 9–3b) should be installed between the line voltage outlet and the power supply. The filter will reduce or eliminate voltage spikes from reaching the workpiece via the soldering tool.

Electromagnetic energy exists in a "force field" about certain electrical devices, such as transformers. This electromagnetic energy travels through space and can cause interference with other electrical devices, in the same manner that noise disturbs radio and television programs. The transformer in a power supply can act as a sending station for this electromagnetic energy (Figure 9–3c). The soldering tool an act like an antenna which captures the electromagnetic energy and directs it into the workpiece. Proper grounding of the workpiece to the power supply (Figure 9–3d) can reduce or eliminate this hazard. Grounding by other means may not be successful, since the separate ground of the equipment may have a different voltage.

9–1c GENERAL REPAIR WORK

At various times you may have to repair electronic equipment that has obvious problems, such as cut or frayed line cords, or broken switches, controls, fuse holders, or other parts. These repair jobs are usually straightforward and easy to perform. However, some disassembly is required for most of them.

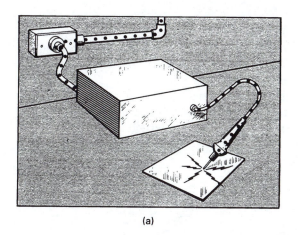

(a)

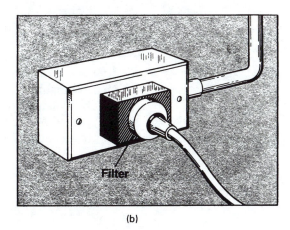

Filter

(b)

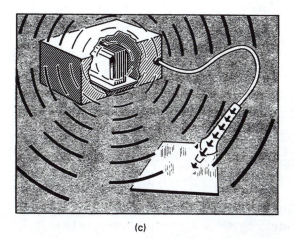

(c)

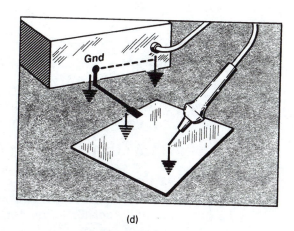

Gnd

(d)

Figure 9-3 Electrical overload problems: (a) typical work situation; (b) filter eliminates power-line spikes; (c) electromagnetic radiation; (d) positive grounding of workpiece. (Permission to reprint granted by PACE Incorporated, 9893 Brewers Court, Laurel, MD, 20707.)

1. Cases or covers usually have to be opened and removed to unsolder or disconnect a line cord. The strain relief will have to be removed and reassembled to pass the new cord into the unit (see Section 5–1c.12 and Figure 5–17).

2. Broken switches have to have the wires unsoldered and then be removed from the chassis or panel (see Section 5–1c.7 and Figure 5–12, and Section 5–1c.8 and Figure 5–13).

3. A fuse holder may be defective and have to be unsoldered and removed from the chassis (see Section 5–1c.10 and Figure 5–15).

4. Control knobs may be broken and have to be replaced simply by pulling them off or loosening a setscrew (see Section 5–1c.6 and Figure 5–11).

9–1d REPAIRING DAMAGED PRINTED CIRCUIT BOARDS

A PC board itself can become damaged. During initial assembly or on a subsequent repair job, the board can become damaged from too much heat and/or pressure applied by the soldering tool. The conductors on the board may lift away from the substrate or be broken completely.

Scratches and nicks in the conductors can reduce the amount of current-carrying area and cause problems in circuit performance or even complete failure. Figure 9–4 shows some examples of damage to circuit conductors.

The board may have a hole in it or a very badly burned area. When too much heat is applied to some substrates, they blister internally and have a spotted appearance referred to as *measling,* which in turn weakens the board and can cause problems with circuit operation.

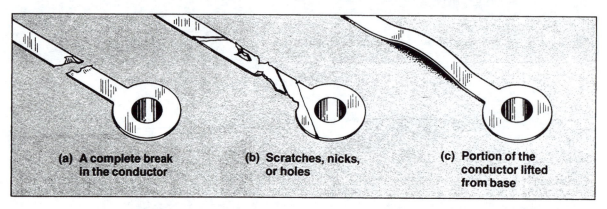

(a) A complete break in the conductor

(b) Scratches, nicks, or holes

(c) Portion of the conductor lifted from base

Figure 9–4 Typical damage to circuit conductors. (Permission to reprint granted by PACE Incorporated, 9893 Brewers Court, Laurel, MD, 20707.)

9–1d.1 Repairing Conductors

A break in a conductor should be trimmed evenly with a very sharp knife. Care must be taken when using the knife so that you do not slip and cut another conductor. The simplest way to repair a conductor is to flatten a piece of tinned wire and place it across the break as shown in Figure 9–5. Each end of the wire is soldered to the original conductor.

Figure 9–5 Surface-mounted jumper wire repair. (Permission to reprint granted by PACE Incorporated, 9893 Brewers Court, Laurel, MD, 20707.)

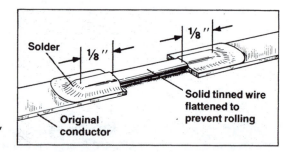

Solder

⅛″

⅛″

Solid tinned wire flattened to prevent rolling

Original conductor

A break in the conductor can be repaired with the through-board jumper wire technique shown in Figure 9–6. A center punch is used to locate holes to be drilled on each of the conductor points. Do not use too much pressure when striking the board. Holes approximately 0.005 in. larger than the wire to be used are now drilled. The wire is formed like the leads of an axial component and pushed through the holes. The ends of the wire are fully clinched into the ends of the conductor. The jumper is now soldered to the conductor. The jumper wire may have some insulation placed on it, which is sometimes referred to as *spaghetti.*

9–1d.2 Repair of Edge Connectors

An *edge connector,* referred to as a *finger,* may lift from the board or be completely missing. Special chemical adhesive is used to reattach the finger to the board. If the finger is not creased or broken, the procedure

Figure 9–6 Through-board method of repair. (Permission to reprint granted by PACE Incorporated, 9893 Brewers Court, Laurel, MD, 20707.)

Original conductor
Solder
1/16″
1/16″
Insulation (optional)
Solid tinned wire

is easy, as shown in Figure 9–7. Chemical activator is applied to both surfaces (Figure 9–7a) and a thin layer of adhesive is applied to the board (Figure 9–7b). The finger is then clamped to the board and held in place by nonadhering Teflon blocks (Figure 9–7c).

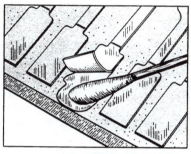
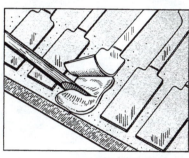

(a) **Applying activator to both surfaces**

(b) **Applying thin layer of adhesive to base material**

(c) **Clamping of copper to base material using nonadhering Teflon blocks**

Figure 9–7 Edge connector repair. (Permission to reprint granted by PACE Incorporated, 9893 Brewers Court, Laurel, MD, 20707.)

9–1d.3 Repair of Substrate

If a part of the substrate of the PC board is damaged, it may have to be removed completely and a new section reconstructed. Figure 9–8 shows one example of repairing the substrate of a board. The damaged area will have to be drilled or cut out. A piece of nonsticking Teflon is taped into place behind the opening (Figure 9–8a). The board is turned over and the hole is filled with a fiberglass–epoxy mixture (Figure 9–8b). After the mixture is cured and hardened it is smoothed with a small abrasive grinding wheel (Figure 9–8c). In the case of a through-hole connection, a small hole will be drilled through the new area. An eyelet or funnelet would be placed into the hole for support and the new conductor soldered to the existing conductor on the board.

The repair techniques that have been given represent a very small sample of the skill and expertise that is required in quality assurance repair of PC boards. For you to develop these advanced skills, it would be necessary to attend a special course given by private schools and companies, such as PACE Incorporated. Often, these organizations will custom-fit a course for the particular repair jobs needed by a company and will conduct the course on the company's own premises.

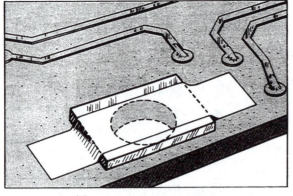

(a) **Hole backed with Teflon**

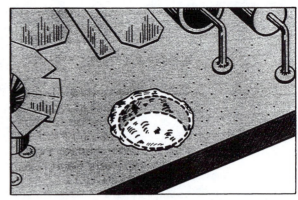

(b) **Hole filled with fiberglass/epoxy mixture**

(c) **Smoothing repair area with abrasive wheel**

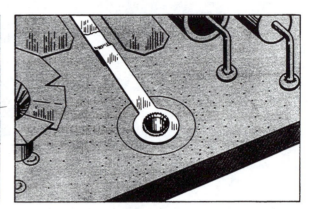

(d) **Reestablishment of circuitry/holes using replacement conductors, pads, and eyelets**

Figure 9-8 Repairing a PC board substrate. (Permission to reprint granted by PACE Incorporated, 9893 Brewers Court, Laurel, MD, 20707.)

≡≡≡ **SECTION 9-2**
DEFINITION EXERCISES

Write a brief description of each of the following terms.

1. State of the art _____

_____

2. Quality assurance _____

_____

3. Rework _____

_____

4. Repair _____

_____

5. Modification _____

_____

6. Salvage _____

_____

7. Conformal coating _____

_____

8. Hard wired _____

_____

9. Measling _____

_____

10. Spaghetti _____

_____

≡ SECTION 9-3
EXERCISES AND PROBLEMS

Complete this section before beginning the next section.

1. List four types of repair jobs.

   a.   **b.**

   c.   **d.**

2. List the four elements of repair.

   a.

   b.

   c.

   d.

3. List the personal attributes of a successful repair person.

   a.

   b.

   c.

   d.

4. List six repair station safety precautions.

   a.

   b.

   c.

   d.

   e.

   f.

5. Explain what causes voltage spikes.

6. Explain the reason for using line filters.

7. Explain electromagnetic radiation and the problems it causes.

8. Briefly describe the importance of grounding a workpiece being repaired.

9. List four general repair jobs that may be performed on electronic equipment.

    a.

    b.

    c.

    d.

10. Fill-in the blanks in the following statements, making them correct, by using the following words: *jumper, heat, epoxy, bonded, pressure.*

    a. One main cause of damaged PC boards where the conductor has

    lifted or the board has measling is too much _____ and

    _____ from the soldering device.

b. Conductors on a PC board can be repaired by removing the damaged section and soldering a _____ across it.

c. Edge connectors may have to be replaced on a PC board and then _____ to the substrate with special chemicals.

d. The substrate of a board may be damaged and the area has to be removed and filled with _____ and then sanded smooth.

≣ SECTION 9–4
EXPERIMENT

EXPERIMENT 1. Repairing Electronic Equipment

Objective:

To gain skills in repair work.

Introduction:

This particular experiment depends on obtaining electronic equipment and PC boards that require simple repairs.

Materials Needed:

Basic electronic hand tools
Soldering iron and desoldering tool (see Section 9–1a.4)

Procedure:

1. Locate the repair to be done.
2. Refer to the proper section in this unit for the type of repair to be accomplished.

Fill-in Questions:

1. To repair an electric device or PC board, you must have the _____ tools.

2. The four personal attributes for successful repair are, _____ , _____ , _____ and _____ .

≣ SECTION 9–5
INSTANT REVIEW

- The four *types of repair* are rework, repair, modification, and salvage.
- The five *elements of repair* are the skill of the repair person, proper tools and equipment, analysis and definition of the problem, defined procedures of work, and quality assurance.
- *Personal attributes for successful repair* are skills, responsibility, patience, and confidence.
- *Proper repair equipment* includes hand tools, soldering tools, desoldering tools, special equipment, and miscellaneous supplies.
- *Repair station safety precautions* are that:
  - Tools be in good working condition
  - The work area be well lighted
  - The work area be clean and free of debris
  - Safety glasses be worn during repair procedures
  - Caution and safety signs be displayed
  - A first-aid kit be available
  - Electrical line filters be installed
  - The workpiece and equipment be properly grounded
  - There is an antistatic tabletop and grounding wrist strap

- *General repair work* includes the replacement of broken or defective switches, fuse holders, control knobs, line cords, and other obvious problems.
- *Conductors* on a board may become damaged. A clean cut is made and then a metallic jumper is soldered across the two pieces of conductor.
- *Edge connectors* may lift off the board and have to be rebonded with a special adhesive.
- The *substrate* of a board may be damaged and have to be drilled or cut out. Fiberglass and epoxy are then used to restore the board to its original condition.

≡≡≡ SECTION 9-6
SELF-CHECKING QUIZ

Circle the most correct answer for each question

1. Updating an assembly to meet new specifications is called:
   a. rework
   b. modification
   c. hard wired
   d. none of the above

2. One method of reducing or eliminating voltage spikes at the workstation is to:
   a. use an antistatic tabletop
   b. use a grounding wrist strap
   c. use a line filter
   d. wear rubber sole shoes

3. One method of reducing electromagnetic radiation at the workstation is to:
   a. ground the workpiece
   b. use a line filter
   c. wear a grounding wrist strap
   d. unplug the soldering/desoldering tool after each operation

4. When repairing MOS devices you should:
   a. have the workpiece grounded
   b. wear a grounding wrist strap
   c. use a line filter
   d. all of the above

5. A jumper wire on a PC board may have insulation on it called:
   a. conformal coating
   b. spaghetti
   c. measling
   d. none of the above

6. Which of the following is *not* a good attribute for a successful repair person?
   a. skills
   b. responsibility
   c. expediency
   d. confidence

7. To repair a through-hole connection in a PC board, you would use:
   a. an eyelet
   b. a funnelet
   c. both a and b
   d. neither a nor b

8. Unsoldering components from a PC board to be used again is referred to as:
   a. rework
   b. hard wired
   c. salvage
   d. modification

9. Measling is a term referred to when:
   a. a PC board develops bubbles under the surface of the substrate from too much heat
   b. the solder boils and forms small bumps on the joint
   c. the rosin boils and leaves a residue on the board
   d. none of the above

10. The repair of a PC board should be performed with the same high-quality techniques as those used to produce the original piece.
    a. True
    b. False

ANSWERS TO FILL-IN QUESTIONS AND SELF-CHECKING QUIZ

Experiment 1:   **(1)** proper   **(2)** skills, responsibility, patience, confidence.

Self-Checking Quiz: **(1)** b   **(2)** c   **(3)** a   **(4)** d   **(5)** b   **(6)** c   **(7)** b   **(8)** c   **(9)** a   **(10)** a

Unit 10

Automated Electronic Assembly

INTRODUCTION

Several steps are required in electronics assembly before a unit is completed. In the past, many of these steps, such as inserting components into a PC board and soldering them, required many people. Today, however, some manufacturers may use machines for some or all of the assembly procedures.

Production-line assembly systems have been in use for some time in the manufacture of automobiles, home appliances, and other products. A *production line* is a means of producing a product or subunit for a product with the most efficient methods and in the shortest possible time. The device or unit is produced in stages at specific stations or work locations. A *conveyor system* of rollers, belts, chains, and motors moves each unit from one workstation to the next in a progressive, predetermined manner.

Printed circuit boards are particularly suited for this type of production.

UNIT OBJECTIVES

Upon completion of this unit, you will be able to:

1. Describe how PC boards are assembled.
2. Explain how wave soldering is accomplished.
3. List the various machines used with an automatic insertion mount component PC board assembly.
4. Explain the use of surface mount component machines used in automatic PC board assembly.
5. Identify existing jobs that the electronic assembly person can cross-train into and new job possibilities with the ever-changing technology.
6. Define various terms used in the manufacturing process of electronic products.

An *assembly line* is a production line that usually involves only assembly work. Production lines may include painting, packaging, and stacking.

10-1a   General PC Board Assembly Line

Generally, certain basic steps are used in the assembly of PC boards on an assembly line, as shown in the block diagram of Figure 10-1. All of the parts used on a PC board must be received before work is begun on the assembly line. These parts first enter the shipping and receiving department of the company. From here the parts are sorted, categorized, and placed in appropriate places in the stockroom. If the blank PC boards have been manufactured by a subcontractor (another company that makes only the board), they must also be stored in the stockroom. The main company may fabricate its own PC board in-house (at the same facility). In either case, the blank PC board is placed on the conveyor system and heads for the first assembly station.

There may be one or more stations where preassigned components are inserted into the board. These stations will have bins or rolls of components with the leads trimmed to the exact length.

After all of the components are inserted into the board, it moves along the conveyor system to the wave soldering machine, which solders all joints simultaneously. Once the joints cool down they are cleaned with a solvent to remove excess flux.

The completed PC board is inspected and tested in the next step of the operation. If a board does not meet the test criteria, it will go to a rework station. After components have been replaced and soldered by hand, the board will be reinspected. When the PC board passes inspection it may be sent to the shipping and receiving department for another destination or it may go to the finished stockroom.

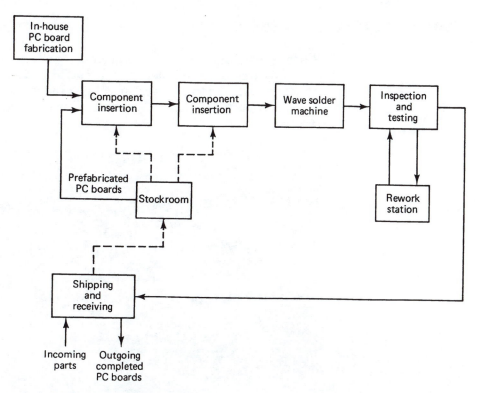

Figure 10–1   General block diagram for PC board production line.

10–1b   Wave Soldering

Wave soldering is an automatic process of soldering components on a PC board as shown in Figure 10–2. The *waves* of *flux* and *solder* are produced by pump motors which cause the fluid to rise slightly above the container edge in the center. This wave contacts the underside of the board and component leads.

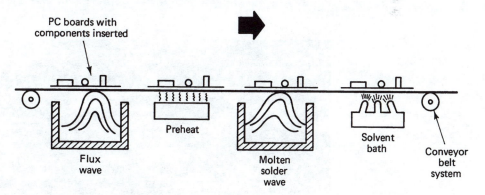

Figure 10–2   Wave soldering.

The PC boards with the components inserted move along the conveyor system from left to right. First, flux is applied by the flux wave. Next, the board is preheated so that the dwell time of the molten solder is kept to a minimum to protect the components from overheating. The board then passes over the solder wave, where the joints are soldered. Finally, a solvent bath removes any excess flux from the board. The board is then cooled and moves on into the inspection station.

Various machines can be connected together to form an automated assembly line. In a completely automated system the PC board travels from one machine to another on conveyor systems without human intervention. Figure 10–3 shows a complete insertion mount technology (IMT) system.

Figure 10–3 IMT assembly system. (Permission to reprint granted by Universal Instruments Corporation, Box 825, Binghamton, NY 13902.)

An IMT system handles regular components such as resistors, capacitors, transistors, and ICs. Figure 10–4 shows some components and PC boards of this system. Notice how the components are attached to a paper stringer that feeds into the insertion machines.

Figure 10–4 IMT PC boards and components. (Permission to reprint granted by Universal Instruments Corporation, Box 825, Binghamton, NY 13902.)

Figure 10–5 shows a closer view of components mounted in a PC board. How many of these components can you identify? A good assembly person must know how to recognize most of the components used on various PC boards.

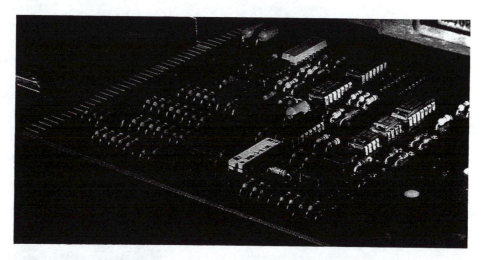

Figure 10–5   IMT components mounted in PC board. (Permission to reprint granted by Universal Instruments Corporation, Box 825, Binghamton, NY 13902.)

10–1c.1   Insertion Mount Component Machines

In this section we show some examples of component insertion machines.

10–1c.1 a   Axial-Lead Components

Figure 10–6 shows an axial lead insertion machine placing a component into a PC board. Figure 10–7 shows how the components are attached to the paper stringer and feed into the machine.

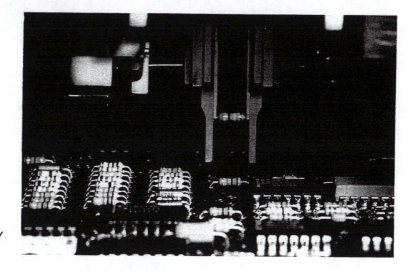

Figure 10–6   Axial-lead component insertion. (Permission to reprint granted by Universal Instruments Corporation, Box 825, Binghamton, NY 13902.)

10–1c.1 b   Radial-Lead Components

Figure 10–8 shows a radial-lead insertion machine placing a component into a PC board.

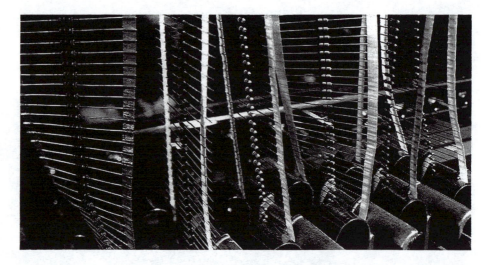

Figure 10–7 Axial-lead component strips. (Permission to reprint granted by Universal Instruments Corporation, Box 825, Binghamton, NY 13902.)

Figure 10–8 Radial-lead component insertion. (Permission to reprint granted by Universal Instruments Corporation, Box 825, Binghamton, NY 13902.)

10-1c.1 c   Dual-in-Line Components

Figure 10–9 shows a DIP insertion machine placing an IC into a PC board. The DIP ICs are packaged in special plastic or metal carriers by the component manufacturer, as shown in Figure 10–10. These carriers protect the ICs during shipment and also serve as a holder for them as they are fed into the insertion machine.

10-1c.1 d   Robotics for Assembly

The insertion machines mentioned previously perform only one type of mechanical action for a specific type of component. They are a limited form of what is called robotics. *Robotics* is the use of machines that are programmed by a computer to perform mechanical operations similar to those that might be performed by a human being. The most popular robot today simulates the action of the human arm and its parts have similar identifying names, such as shoulder, wrist, and elbow. An industry standard of the robotic arm is called the *SCARA robot*, shown in Figure 10–11. This robot can be programmed to insert axial-, radial-,

Figure 10–9 DIP component insertion. (Permission to reprint granted by Universal Instruments Corporation, Box 825, Binghamton, NY 13902.)

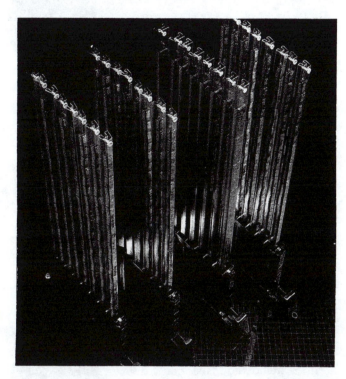

Figure 10–10 DIP carrier packages. (Permission to reprint granted by Universal Instruments Corporation, Box 825, Binghamton, NY 13902.)

Figure 10–11 SCARA robot used in assembly. (Permission to reprint granted by Universal Instruments Corporation, Box 825, Binghamton, NY 13902.)

and DIP-lead components and to perform other operations on the same PC board within the same time frame.

10-1c.1e *Automatic PC Board Inspection*

Within an automated system the inspection of PC boards is also done automatically. The inspection of each board is accomplished with a special camera lens, closed-circuit television system, and computer system. Figure 10–12 shows an automatic PC board inspection system, which is usually part of the overall main automated assembly system.

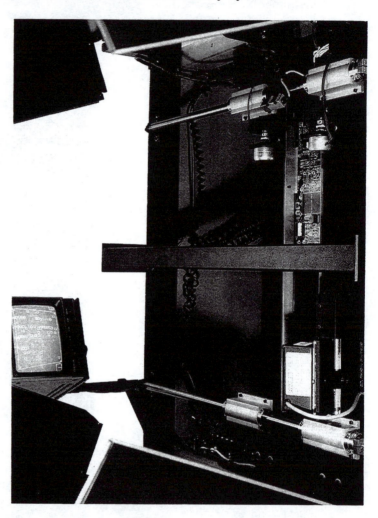

Figure 10–12 Automatic PC board inspection system. (Permission to reprint granted by Universal Instruments Corporation, Box 825, Binghamton, NY 13902.)

10–1d SURFACE MOUNT TECHNOLOGY ASSEMBLY

Surface mount components (SMCs) require different types of machines for assembling PC boards. A complete SMC automated assembly system is shown in Figure 10–13. A SMC PC board will have a slightly different appearance than that of an IMT board. Figure 10–14 shows a SMC board with an IC being placed by a machine.

10–1d.1 PC Board Printing System

In a SMC system the PC board may have the foil pattern printed on it as part of the operation. A solder flux paste is also applied to the conductors to help hold the components on the board as the board goes through the soldering process. Figure 10–15 shows a SMC pass-through printing system.

Figure 10–13 SMC automated assembly system. (Permission to reprint granted by Universal Instruments Corporation, Box 825, Binghamton, NY 13902.)

Figure 10–14 SMC PC board. (Permission to reprint granted by Universal Instruments Corporation, Box 825, Binghamton, NY 13902.)

Figure 10–15 SMC pass-through printing system. (Permission to reprint granted by Universal Instruments Corporation, Box 825, Binghamton, NY 13902.)

10-1d.2 SMC Mounting Machine

The SMC mounting machine uses a nozzle with a vacuum to pick up and place the component at the proper location on the PC board. This operation is very exacting and is performed with the aid of a computer. Figure 10-16 shows a SMC mounting machine placing an IC on a PC board.

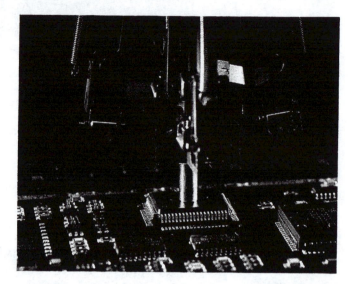

Figure 10-16 SMC mounting on PC board. (Permission to reprint granted by Universal Instruments Corporation, Box 825, Binghamton, NY 13902.)

10-e OPPORTUNITIES FOR EMPLOYMENT

Automated electronic assembly systems replace many assembly jobs formerly done by human beings. Perhaps two or three people are required to operate an automated system. These persons will be responsible for minor tasks, such as keeping components reaching the machines, checking the continuous flow of boards along the system, and other operations. A former electronics assembler can handle most of these jobs, but the position will have many applicants and be very competitive. A good contestant for a position on an automated system will have skills and knowledge of every aspect of electronic assembly. This person will have to study more about overall automated systems and continue to upgrade the skills required in the ever-changing electronics world.

SECTION 10-2
DEFINITION EXERCISES

Write a brief description of each of the following terms.

1. Production line _____

_____

2. Conveyor system _____

_____

3. Assembly line _____

_____

4. Subcontractor _____

_____

5. In-house _____

_____

6. Wave soldering _____

_____

7. Flux wave _____

_____

8. Solder wave _____

_____

9. Robotics _____

_____

10. SCARA robot _____

_____

≡≡≡ **SECTION 10–3**
   EXERCISES AND PROBLEMS

Complete this section before beginning the next section.

1. List the general procedure for PC board production (refer to Figure 10–1).

   a.

   b.

   c.

   d.

   e.

   f.

   g.

   h.

2. Describe the process of wave soldering (refer to Figure 10–2).

3. Explain the difference between IMT and SMC automated machines.

4. Why would a SCARA robot be used instead of regular insertion machines? Explain.

5. How are ICs placed on a PC board with SMC machines?

6. List some components or devices used in an automated inspection system.

7. List some jobs or employment opportunities that an electronics assembler might perform with an automated electronics assembly system.

≡≡≡ **SECTION 10-4**
EXPERIMENT

The experiment for this section cannot be performed since no actual manual skills are involved. However, you should visit a manufacturing plant that has some automated equipment. By actually seeing this type of equipment in operation, you can gain more knowledge and understanding of its use. Ask your instructor to arrange a field trip to a facility that uses some automated equipment.

Use the following form to save information about the trip.

Name of company _____

Address _____

City and state _____

Company telephone number _____

Nature of business or product _____

Date of trip _____ Time of trip _____

to _____

Person contacted at company _____

Mode of transportation _____

Comments or discussion after trip:

- An *assembly line* is a production line that usually involves only assembly work.
- The *general procedure for PC board production* involves:
  1. Receiving the parts
  2. Storing the parts in the stockroom
  3. Bringing the boards and parts together for insertion and mounting
  4. Wave soldering the boards
  5. Inspecting and testing the boards
  6. Reworking any boards that do not pass inspection and testing
  7. Sending reworked boards back to inspection and testing
  8. Preparing completed boards for shipping or installation in equipment
- *Wave soldering* is an automatic process of soldering components on a PC board simultaneously.
- *Automated assembly lines* have machines that insert or mount components, solder, inspect, and package PC boards.
- *IMT* stands for "insertion mount technology."
- *SMC* stands for "surface mount components."
- Components such as *resistors* and *capacitors* are attached to paper stringers that are fed into insertion machines.
- *DIP ICs* are packaged in special plastic or metal carriers that are placed directly on insertion machines for use.
- *Insertion machines* are a form of robotics.
- The *SCARA robot* can perform several tasks on the same PC board, such as inserting axial lead components, radial lead components, or DIP ICs, and other operations.
- In an *automated assembly system* the inspection of boards is performed automatically with video cameras, closed-circuit television, and computers.
- With a *SMC assembly system* the conductor patterns of the boards might also be printed with the system.
- A *solder flux paste* is applied to SMC boards to help hold the components to the board while they are being soldered.
- An *electronic assembly person* can still find employment in an automated assembly line by updating his or her skills and studying the operations required in the system.

Circle the most correct answer for each question.

1. An assembly line is where:

   a. people gather for instructions

   b. products are mass produced

   c. parts are stored

   d. none of the above

2. The process in which entire PC boards are soldered at one time is referred to as:

   a. contact soldering

   b. wave soldering

   c. solid soldering

   d. all of the above

3. PC boards are moved from one work location to another by:

   a. an assembly line

   b. a production line

   c. a conveyor system

   d. slides

4. If a particular PC board does not pass inspection, it may be sent to the station called:

   a. component insertion
   b. wave solder

   c. reject
   d. rework

5. The basic PC board without components may be manufactured:

   a. in-house

   b. by a subcontractor

   c. both a and b

   d. neither a nor b

6. The type of automatic machine to use for axial lead components is:

   a. IMT
   b. SMC

   c. both a and b
   d. neither a nor b

7. The components placed into an insertion mount machine must be loaded by hand, one by one.

   a. True
   b. False

8. The ICs placed on boards using SMC machines is accomplished by the use of:

   a. magnetism
   b. a gripper

   c. a vacuum
   d. all of the above

9. A machine that can perform many tasks on a single PC board is:

   a. an IMT machine
   b. an SMC machine

   c. an automated machine
   d. a SCARA robot

10. There are no jobs available for assembly persons on an automated assembly line.

    a. True
    b. False

ANSWERS TO SELF-CHECKING QUIZ

(1) b (2) b (3) c (4) d (5) c (6) a (7) b (8) c
(9) d (10) b

Appendix A

Miscellaneous Resistor Color Coding

The coding of each type of resistor shown in Figure A–1 is indicated by numbers and letters with arrows:

1 = first significant figure   M = multiplier
2 = second significant figure   T = tolerance

Figure A–1 Miscellaneous resistor color-code types.

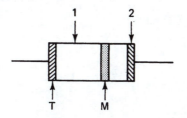

Body-end-band type   Body-end-dot type

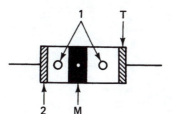
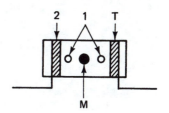

Dot-band types

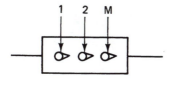

Body-dot type

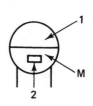

Miniature type

Appendix B

Miscellaneous Capacitor Color Coding

The coding of each type of capacitor shown in Figure B–1 is indicated by numbers and letters with arrows:

 1 = first significant number
 2 = second significant number
 M = multiplier
 T = tolerance
 C = class or characteristic
TC = temperature coefficient
CI = code identification (either EIA or JAN)
 V = voltage rating
 W = wide white band to distinguish capacitor from resistor

The color code values for capacitors are shown in Table B–1.

TABLE B–1

General Color Code For Capacitors

Color	Capacitance 1st and 2nd Number	Multiplier	Capacitance Tolerance	Temperature Range	Working Voltage
Black	0	1	±20%	−55°-+70°C	
Brown	1	10	±01%		100 (EIA)
Red	2	100	±02%	−55°-+85°C	
Orange	3	1000			300
Yellow	4	10,000		−55°-+125°C	
Green	5	100K	±05%		500
Blue	6	1M		−55°-+150°C	
Violet	7				
Gray	8				
White	9				
Gold		0.1	±0.5%		1000 (EIA)
Silver		0.01	±10%		

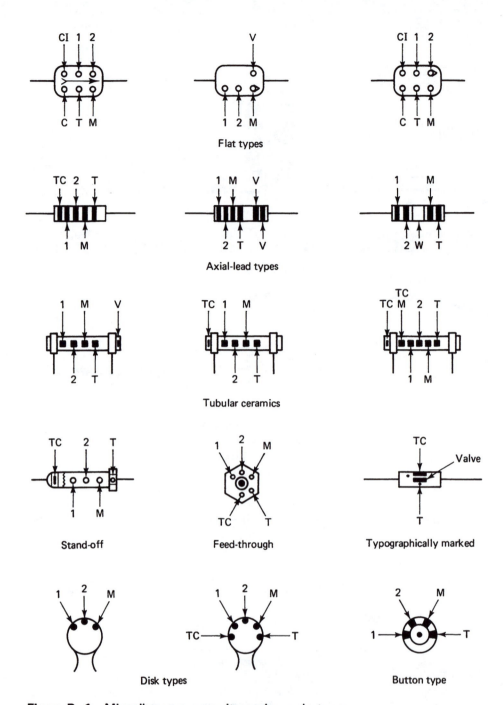

Figure B–1 Miscellaneous capacitor color-code types.

Appendix B / Miscellaneous Capacitor Color Coding

Appendix C

Interpreting Number/Letter-Coded Capacitor Values

Numbers and letters are used on some disk capacitors to indicate their value, as shown in Figure C–1. Table C–1 shows the assigned values. For example, the indication:

$$103K = 1 - 0 - 000 = 10000 \text{ pF} = 0.01 \ \mu F \text{ at } 10\%$$

and

$$472J = 4 - 7 - 00 = 4700 \text{ pF} = 0.0047 \ \mu F \text{ at } 5\%$$

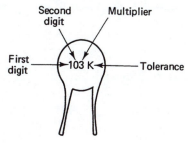

Figure C–1 Number/letter code capacitor identification.

TABLE C–1
Number/Letter Coded Capacitor Values

Digit	Multiplier	Tolerance 10 pF or less	Tolerance Over 10 pF	Letter
0	1			
1	10	±0.1 pF		B
2	100	±0.25 pF		C
3	1000	±0.5 pF		D
4	10,000	±1.0 pF	± 1%	F
5	100,000	±2.0 pF	± 2%	G
—			± 3%	H
—			± 5%	J
8	0.01		±10%	K
9	0.1		±20%	M

Index

A

AMP cable connector, 235
Active device, 73
Adjustable resistor, 36–37
Adjustable wrench, 107–08, 129
Allen wrench, 109–10, 129
Alligator clip, 135, 156, 169
Ammeter, 13, 27
Ampere, 1, 3, 26
Amplifier, 76, 80, 85
Antenna, 51, 69
Antiwicking tool, 169, 186
Armature, 48, 69
Assembly line, 266, 277
Audio, 37, 49–50, 69, 80
Automated, 268, 277
Axial leads, 42, 68, 195–96, 198–200,
    215, 225, 269–701

B

Ball peen hammer, 104–05
Bar folder, 118
Bench vise, 105–06, 129
Binding post, 51, 69
Bipolar transistor, 76, 85, 99, 202
Blow hole, 44, 68
Box-end wrench, 107–08, 129
Brakes, 118
Breakout, 233, 245
Bristo wrench, 109, 130
Buzzer, 50, 69

C

C-clamp, 105–06, 129
Cable clamp, 134–35, 143, 156
Cable lacing, 232, 245
Cap nut, 141, 156
Capacitor, 31, 42–44, 68
Capillary action, 172, 186

Carbon resistor, 32–34, 67
Carriage bolt, 136–37, 155
Channel-lock pliers, 105, 129
Chassis, 80, 99, 134, 155
Chassis ground, 83, 100
Chassis mount nut, 143, 156
Chip, 210
Chip components, 210–11, 225
Choke, 46, 68
Chuck key, 114, 130
Circuit breaker, 15, 27
Cladding, 225
Claw hammer, 104–05, 129
Clinched, 197–99, 225
Closed circuit, 1, 3, 27
Cold solder joint, 167, 186, 200, 225
Color code, 31–33, 35, 68
Component mounting terminal, 135,
    156
Conductive inks, 213, 226
Conductors, 1, 4, 191, 225, 251, 255–
    56, 262
Conformal coating, 214, 226, 251
Control knob, 135, 145, 156
Conveyor system, 265–66
Copper foil, 190
Copper pattern, 190, 191
Countersunk hole, 137–38, 156
Crescent wrench (*See* Adjustable
    wrench)
Crystal (*See* Piezoelectric crystal)
Current, 1, 3, 26, 74, 76–78, 82, 99

D

DIAC, 78
DIP, 81, 100, 203–05, 209, 213, 215,
    239, 251, 270–71, 277
DPDT switch, 9–10, 27
DPST switch, 9–10, 27
Deposited-film resistor, 35, 68

Desoldering bulb, 171, 186
Desoldering tool, 172, 186, 214–16,
    226
Diagonal cutters, 113–14, 130, 252
Digital module IC, 80, 99
Diodes, 73–76, 99
Double-sided board, 194–96, 225
Drill bits, 114–15, 130
Drill press, 117, 130
Dwell time, 169, 186

E

Earth ground, 16, 27, 83, 100
Edge connector, 84, 191, 256, 262
Electrical circuit, 1–3, 26
Electrical conductor, 1–2, 4, 26
Electrical hardware, 133, 155
Electrical insulator, 1, 9, 26
Electrical jack, 7–8, 26
Electrical plug, 7–8, 26
Electrical shock, 16–17, 27
Electrical switch, 9–11, 27
Electrically conductive adhesives,
    212–13, 226
Electrician's 6-in-1 tool, 111–12, 130
Electrocution, 16–17, 27
Electrolytic capacitor, 42–44, 68
Electronics assembler, 103, 274
Etchant, 192
Eutectic solder, 161, 185
Eyelets, 193, 225, 257

F

Farad, 42
Female connector, 84, 100
Fiber washer, 134–35, 156
Flat pack, 205–07
Flat washer, 134–35, 143–45, 156
Flat wire, 6–7, 229, 236, 245
Flat-blade screwdriver, 109, 129, 252

Flexible board, 194, 225
Flip-flops, 80–81
Flux, 159, 225, 251
Flux wave, 267
Fractured solder joint, 167, 186
Frayed, 19, 27, 116
Free electrons, 4
Funnelets, 193, 225, 257
Fuse, 14

G

Ganged, 9, 27
Gates, 80–81, 100
Grinder, 117–18, 130
Grommet, 134–35, 143–44, 156
Ground plane, 191
Gull-wing leads, 226

H

Hacksaw, 104–05, 129
Hand drill, 114–15, 130
Hard wired, 216
Hardware, 133, 155
Harness board, 233
Heat cycle, 168, 186
Heat recognition time, 169, 186
Heat sink, 78–80, 99, 169, 190, 225
Heat-shrinkable tubing, 170–71, 186, 197, 230–31
Helical potentiometer, 38, 68
Henry, 46
Hex nut, 134–35, 156
Hex wrench, *See* Allen wrench
High packing density, 189
Hole cutter, 114, 130
Hybrid IC, 80, 99

I

In-house, 266
Incandescent lamp, 11, 27
Indium, 161–62, 185
Inductor, 31, 45–46, 68, 211
Insertion mount technology (IMT), 205, 213, 225, 268–69, 277
Integrated circuit, 35, 80
Interference nut, 141, 156

J

J-lead, 209, 226
JFET, 77, 85, 99
Jigsaw, 116–17, 130
Junction Field-Effect Transistor, *See* JFET

K

Key, 234
Kilo, 31, 40–41
Knurled thumb nut, 141, 156

L

LASCR, 82
LCCC, 209–10, 225
LED, 11–12, 27, 202
Laminate, 191, 225
Lap flow tool, 207, 226
Linear module IC, 80, 99
Linen cord, 230, 232, 245

Load, 3, 18, 27
Logic symbol, 80, 85, 100
Long nose pliers, 114, 130, 197, 252
Loose tubing, 231, 245
Loudspeaker, 50, 69, 78

M

MOSFET, 77, 85, 99
Machine bolt, 136–37, 155
Male connector, 84, 100
Measling, 256
Mega, 31, 40–41
Metal files, 110–11, 130, 252
Meter, 50, 69
Mica, 80, 99
Micro, 31, 40–41
Microphone, 49–50, 69
Milli, 31, 40–41
Modification, 250, 261
Monolithic IC, 80, 99
Multidigit display, 50, 69
Multilayer board, 194–95, 225
Multimeter, 1, 12–13

N

NPN transistor, 76, 99
Neon lamp, 11–12, 27
Normally closed contacts, 9–10, 27, 48–49, 69
Normally open contacts, 9–10, 27, 48, 69
Nylon cable ties, 230–31, 245

O

Ohm, 3, 12–13
Ohmmeter, 1, 12, 26, 31, 73
Op amp, 81, 85
Open circuit, 3, 5
Open-end wrench, 107–08, 129
Optoelectronics, 82, 100
Optoisolator, 83, 100
Overheated joint, 167, 200–201, 225
Overload, 14, 18, 27, 254–55
Oxide, 162, 164, 185, 191

P

PC board heat sink, 190–91
PLCC, 209, 225
PNP transistor, 76, 99
Pads, 191, 225, 251
Passive device, 31
Phillips-head screwdriver, 109, 129, 252
Photodector, 82, 100
Photoresist, 191
Photoresistor, 39
Photosource, 82, 100
Phototransistor, 82
Pico, 40–41
Piezoelectric crystal, 50, 69
Planar-mounted, 205–06
Potentiometer, 31, 36–38, 68
Power rating, 34, 68
Power transistor, 78–79, 99
Powers of ten, 40–41
Primary winding, 46, 69
Printed circuit board, 189–193

Production line, 265, 267
Prototype, 233, 245
Punches, 114–15, 118

Q

Quality assurance, 250–51, 261

R

Radial leads, 42, 68, 201–02, 215, 269–70
Ratchet wrench, 108, 129
Rectifiers, 74, 76, 99
Reflow solder flux, 212
Reflow solder technique, 189, 207, 226
Reinforced hole, 193, 225
Relative thermal mass, 168, 185
Relay, 46–49, 78, 211
Repair, 249–57, 261
Resistance, 1, 3–6, 12–13, 26, 82
Rework, 250, 261
Rheostat, 38, 68
Ribbon cable, 7, 229, 236
Robotics, 270–71
Rosin joint, 167, 200, 225
Rosin-core solder, 162–63, 185, 252
Runs, 191, 251, 255

S

SCARA robot, 270, 277
SCR, 78
SMC, 205, 225, 272–73, 277
SMD, 205, 225–26, 253
SMT, 210–12, 225, 252
SOIC, 208, 226
SOT, 207–08, 226
SPDT, 9–10, 27
SPST, 9–10, 27
Salvage, 250–51, 261
Schematic diagram, 3, 73, 83–84, 100
Schematic symbol, 9, 14, 26, 73–77, 80–82
Secondary winding, 46, 69
Self-tapping screw, 134–37, 155
Semiconductor, 73, 76–9
Series circuit, 3
Serrated washer, 134–35
Setscrew, 136–37, 155
Seven-segment display, 50, 69
Shears, 118
Sheet-metal screw, *See* self-tapping screw
Shielded-cable metallic braid, 7, 237, 245
Short circuit, 18
Shrinkable tubing, *See* Heat-shrinkable tubing
Silicon grease, 80, 99
Single-sided board, 193, 225
Slip-joint pliers, 104–05, 107, 129
Slow-blow fuse, 15
Snap-in pin connection, 235
Socket wrench, (*See* Ratchet wrench)
Solar cell, 82, 100
Solder, 159–75, 185, 199–200, 212–13, 267
Solder bridge, 165–66, 186, 199
Solder lug, 134–35, 142–43
Solder lug connection, 234

Solder peaking, 167
Solder sucker, *See* Desoldering tool
Solder wave, *See* Wave soldering
Soldering, 159–60, 162, 185, 225
Soldering aid, 164, 185, 252
Soldering gun, 159, 175, 177, 186
Soldering iron, 159, 163–67, 199, 202, 212–13, 252
Soldering iron holder, 164, 185, 252
Solderless terminals, 134–35
Solid state, 73, 82
Solvent action, 162, 185
Spacer, 135–36, 143, 156
Spaghetti, 256
Speed nut, 141, 156
Splice, 169, 186
Spline wrench, (*See* Bristo wrench)
Split-ring lock washer, 134–35, 142–44, 156
Spot ties, 229, 232–33, 245
Spring nut, (*See* Speed nut)
Square nut, 141, 156
Stand-off, 135, 143, 156
State of the art, 249
Step-down transformer, 46, 69
Step-up transformer, 46, 69
Stop nut, (*See* Interference nut)
Stove bolt, 136–37, 155
Strain relief, 135–36, 148, 156
Subcontractor, 266

Substrate, 80, 99, 191–92, 225, 257, 262
Supported hole, 193, 225
Surface mount technology (SMT), 189, 205, 225, 265, 272–74
Swaging, 193
Switch, 1–4, 9–11, 27, 211, 254–55

T

TRIAC, 78, 82
Temperature control, 168, 186
Template, 233, 245
Terminal points, 83, 100
Thermal linkage, 165, 185
Thermistor, 39, 68
Thermoplastic adhesive, 213, 226
Thermosetting adhesive, 213, 226
Through connection, 193, 225
Through hole, 193, 225, 251, 257
Thyristor, 77–78, 99
Tinning, 164
Transformer, 46, 68
Trimmer capacitor, 43–44, 68
Trimmer resistor, 36–37, 68

U

UJT, 76–77, 99
Unsupported hole, 193, 225
Utility cutter, 110–11, 130

V

Variable capacitor, 44, 68
Vise-grip pliers, 104–05, 129
Volt, 1, 3, 26
Voltage, 1, 3, 14–15, 17, 26
Voltage plane, 191, 225
Voltage surge, 15, 27
Voltage-dependant resistor, 39, 68
Voltmeter, 13, 27

W

Wattage rating, (*See* Power rating)
Wave soldering, 265–66, 277
Wetting, 162, 185
Wicking braid, 171–73, 186, 214
Wing nut, 141, 156
Wire, 1, 3–7, 26–27
Wire cable, 230, 245
Wire de-wrapping tool, 238, 245
Wire harness, 233, 245
Wire stripper, 113, 130, 252
Wire wrapping, 229, 238, 245
Wire wrapping tool, 238, 245
Wire-wound resistor, 34, 67

Z

Zero-ohm resistor, 35, 68
Zipper tubing, 230–31, 245